Introduction to
LINEAR PROGRAMMING
with MATLAB®

Introduction to LINEAR PROGRAMMING with MATLAB®

Shashi Kant Mishra

Bhagwat Ram

CRC Press
Taylor & Francis Group
Boca Raton   London   New York

CRC Press is an imprint of the
Taylor & Francis Group, an **informa** business

A CHAPMAN & HALL BOOK

CRC Press
Taylor & Francis Group
6000 Broken Sound Parkway NW, Suite 300
Boca Raton, FL 33487-2742

© 2018 by Taylor & Francis Group, LLC
CRC Press is an imprint of Taylor & Francis Group, an Informa business

No claim to original U.S. Government works

Printed on acid-free paper
Version Date: 20170726

International Standard Book Number-13: 978-1-138-09226-6 (Hardback)

Library of Congress Cataloging-in-Publication Data

Names: Mishra, Shashi Kant, 1967- author. | Ram, Bhagwat, author.
Title: Introduction to linear programming with MATLAB / Shashi Kant Mishra and Bhagwat Ram.
Description: Boca Raton : Taylor & Francis, CRC Press, 2018. | Includes bibliographical references and index.
Identifiers: LCCN 2017016988| ISBN 9781138092266 (hardback : acid-free paper) | ISBN 9781315104003 (ebook)
Subjects: LCSH: Linear programming--Data processing. | MATLAB.
Classification: LCC T57.74 .M57 2018 | DDC 519.7/2028553--dc23
LC record available at https://lccn.loc.gov/2017016988

Visit the Taylor & Francis Web site at
http://www.taylorandfrancis.com

and the CRC Press Web site at
http://www.crcpress.com

Contents

Foreword

Writing the foreword for this introductory textbook on linear programming with MATLAB by Professors Shashi Kant Mishra and Bhagwat Ram at Banaras Hindu University has brought me back to the memory of the Fall Quarter, 1989, at Stanford University, where the famous Professor George Dantzig was teaching the PhD course, Linear Programming, in the Operations Research Department. It was a four-unit course with a one-hour lab. One of the lab projects was to develop codes for solving linear programming problems. At that time, most of my classmates handed in the codes in Fortran or C++. I remembered that difficult time during the compiling stage, making debugs and correct Do-loops, data structures, etc. using Fortran. Having gone through the project, I learned all the details from theories to the programming parts in linear programming. Later on, there was a period where some scholars were developing convenient software over a spreadsheet environment (like Solver in Excel) for students (especially MBAs) to simply just input the data and hit the return to get sheets of the solution reports. Learning like this may treat the Simplex method as a black box. This may be another extreme way to learn linear programming.

Professors Mishra and Ram write this introductory textbook in a clever way; with very light background in linear algebra and MATLAB, the students will be brought to the theory parts quickly. Friendly examples are given to illustrate the theory sections, and MATLAB codes are provided to demonstrate the results. MATLAB is useful here because of its interpreter feature, which allows students to verify step-by-step in the simplex method without the need of compiling the codes. The authors also provide convenient "functions", which are the main steps in the simplex method. Students can simply call the functions to implement some steps in the simplex methods. In this way, the Simplex method is no longer a black box for our students. For the undergraduates, the authors make a very nice trade-off among learning theories, coding parts, and self-assessment of understanding the subject.

Linear programming has long been recognized with beautiful theories as well as wide applications in the practical world. Using MATLAB gives students the chance to "learn by doing", one of the effective learning strategies emphasized in our modern education, in assessing themselves the level of understanding of the linear programming subject. We strongly believe that students who learn the linear programming with MATLAB will definitely understand the subject much better in theories and practical applications.

Sy-Ming Guu
Professor,
Graduate Institute of Business and Management
Dean, College of Management,
Chang Gung University,
Taoyuan, Taiwan
Ph.D. in Operations Research,
Stanford University

Preface

George B. Dantzig formulated a linear programming problem and developed the simplex method to solve it. This new mathematical technique found a wide range of practical applications. This is an introductory textbook on linear programming with MATLAB®, written mainly for students of mathematics, computer science, engineering, economics, management science and agriculture. The textbook is based on the lecture notes and experience of the first author while teaching mathematics Bachelor of Science students at the Banaras Hindu University, Varanasi, India for several years. A large number of available textbooks have been a source of inspiration for introduction of concepts and problems. We are thankful to the authors of those books for their indirect help.

There are many textbooks on linear programming but very few on linear programming with MATLAB. Moreover, among the available textbooks on linear programming with MATLAB, there is a lack of student-friendly textbooks. There was a desperate need of a textbook on linear programming with MATLAB for the beginner of such a course. The purpose of this textbook is to introduce linear programming and use of MATLAB in the formulation, solutions and interpretation of linear programming problems in a natural way. The textbook has been written in a simple and lucid language so that a beginner can learn the subject easily. A prerequisite is a standard single-variable calculus and introductory linear algebra course. Although some background knowledge of multivariable calculus and some experience with formal proof writing are helpful, these are by no means essential.

The textbook has been organized in nine chapters. The first three chapters are an introduction, background of linear algebra needed in the sequel and basic knowledge on MATLAB. Chapter 4 is on simple examples of linear programming problems, concept of convex sets and graphical solution of linear programming problems. Chapters 5 and 6 are on Simplex method with illustrative examples that are solved manually and several examples are solved using MATLAB. Chapter 7 is on duality results and dual simplex method, and the last two chapters are on transportation and assignment problems with a sufficient number of examples. A good number of suitable exercises is also given on each method and with answers at the end of textbook. The textbook contains 80 solved examples to illustrate various methods and applications, and out of these, 42 examples are solved manually and 38 examples are solved using MATLAB.

We have written 18 user-friendly functions which show the step-by-step solution of linear programming problems. This will be an effective concept to those learners who want to learn the programming concept in linear programming.

We are thankful to Prof. Niclas Borlin, Department of Computing Science, Ume University, Sweden who permitted us to use his MATLAB function: hungarian.m. We are also thankful to Senior Acquisitions Editor of CRC, Mrs. Aastha Sharma, for guiding us during the development of this book in LaTex.

Shashi Kant Mishra
Bhagwat Ram
Banaras Hindu University,
Varanasi, India

List of Figures

List of Tables

Chapter 1

Introduction

1.1 History of Linear Programming

We are presenting a theory whose official birth was at the heart of the twentieth century and in fact in the years right after the Second World War. However, all the readers are familiar with the method of Lagrange multipliers from Calculus, named after Joseph Louis Lagrange (1736–1813) who considered equality constrained minimization and maximization problems in 1788, in the course of the study of a stable equilibrium for a mechanical system.

FIGURE 1.1: J. L. Lagrange (1736–1813)

FIGURE 1.2: Joseph B. Fourier (1768–1830)

The famous French mathematician Joseph B. Fourier (1768–1830) considered mechanical systems subject to inequality constraints, in 1798, though Fourier died before he could raise any real interest of his new findings to the mathematical community. Two students of Fourier—the famous mathematician, Navier, in 1825, and the equally famous mathematical economist, Cournot, in 1827, without mentioning the work of Fourier—rediscovered the principle of Fourier, giving the necessary conditions for equilibrium with ad hoc argument which make specific reference to the mechanical interpretation.

In 1838, the Russian mathematician Mikhail Ostrogradsky (1801–1862) gave the same treatment in the more general terms. He asserted without referring to Joseph B. Fourier, that at the minimizer the gradient of the objective function can be represented as a linear combination, with nonnegative multipliers of the gradients of the constraints.

It is worth noticing that Ostrogradsky was a student in Paris before he went to St. Petersburg, and he attended the mathematical courses of Fourier, Poisson, Chauchy and other famous French mathematicians.

FIGURE 1.3: Mikhail Ostrogradsky (1801–1862)

The Hungarian theoretical physicist Julius Farkas (1847–1930) focused on the mathematical foundation and developed a theory of homogeneous linear inequalities which was published in 1901. However, the first effective acknowledgment of the importance of the work of Farkas was given in the Masters thesis of Motzkin in 1933. But, the Farkas Lemma has to wait almost half a century to be applied. American mathematicians also started developing a theory for systems of linear inequalities followed by a paper on "preferential voting" published in *The American Mathematical Monthly* in 1916.

FIGURE 1.4: Julius Farkas (1847–1930)

Note that the theory of linear programming did not just appear overnight. Linear programming depends on development of other mathematical theories and mathematical tools, one of these is of course Convex Analysis, which was not known well before. The birth of the linear programming theory took place in two different, equally developed countries: the USSR and USA, but the motivating forces were also entirely different.

In the USSR, the father of linear programming is Leonid Vitalievich Kantorovich (1912–1996) and he is well known in the mathematical community for his achievements in linear programming, mathematical economics and functional analysis. He was awarded the Nobel Prize in 1975 together with T. C. Koopmans (1910–1985).

In the year 1939, Kantorovich was a young professor at the Leningrad University. A state firm that produced plywood and wished to make more efficient use of its machines contacted Kantorovich for a scientific advice. The aim was to increase the production level of five different types of plywood, carried out by eight factories, each with different production capacity. Kantorovich soon realized that this problem has a mathematical structure.

FIGURE 1.5: Leonid Vitalievich Kantorovich (1912–1996)

In 1939, Kantorovich discussed and numerically solved the optimization problem under inequality constraints, in his small book, which was translated to English in 1960. In this book, Kantorovich presented several microeconomic problems from the production planning of certain industries. But, till 1958, economists in the USSR were not in favour to use the theory given by Kantorovich. In 1960, at the Moscow Conference, economists discussed for the first time the use of mathematical methods in economics and planning, and later in 1971 for optimal planning procedures.

FIGURE 1.6: T. C. Koopmans (1910–1985)

The work of Kantorovich was available to the rest of the world in 1960, when Tjalling Carles Koopmans (1910–1985) published an English translation of Kantorovich's work in 1939.

Meanwhile, a similar line of research on inequality constrained optimization took place in the USA independent of the work of the Russians. During the Second World War from 1942 to 1944, Koopmans worked as a statistician at the "Allied Shipping Adjustment Board" and was concerned with some transportation models.

In the same period, George B. Dantzig (1914–2005), who is recognized as the Western Father of Linear Programming, collaborated with the Pentagon as an expert of programming methods, developed with the help of desk calculators. Dantzig finished his studies and became a PhD in mathematics soon after the war ended.

Job opportunities came from the University of California at Berkeley and from the Pentagon. The simplex method discovered by Dantzig to solve a linear programming problem was presented for the first time in the summer of 1947. In June 1947, Dantzig introduced the simplex algorithm to Koopmans who took it to the community of economists namely, K. J. Arrow, P. A. Samuelson, H. Simon, R. Dorfman, L. Hurwiez and others, and the Simplex method

FIGURE 1.7: George B. Dantzig (1914–2005)

became quite a potential method. The Simplex algorithm has been declared as one of the best 10 algorithms with the greatest influence on the development and practice of science and engineering in the twentieth century.

FIGURE 1.8: Cleve Barry Moler (August 17, 1939)

Cleve Barry Moler, the chairman of the Computer Science department at the University of New Mexico, started developing MATLAB in the late 1970s. He designed it to give his undergraduate students for accessing LINPACK (Linear Algebra Subroutines for Vector-Matrix operations) and EISPACK (To compute eigenvalues and eigen vectors) general purpose libraries of algoritms. It soon became popular to other universities also and found a strong interest among the students of applied mathematics. Jack Little and Steve Bangert attracted with this new programming environment and rewrote several developed MATLAB functions in C. Moler, Little and Bangert founded the Mathworks, Inc., in 1984.

MATLAB was first adopted by researchers and practitioners in control engineering, Little's specialty, but quickly spread to many other domains. It is now also used in education for learning and teaching.

Chapter 2

Vector Spaces and Matrices

2.1 Vector

An n vector is a column array of n numbers, denoted as

$$a = \begin{bmatrix} a_1 \\ a_2 \\ \vdots \\ a_n \end{bmatrix}. \tag{2.1}$$

The number a_i is called the ith component of the vector a. For example, $a = \begin{bmatrix} 1 \\ 2 \\ -3 \end{bmatrix}$ is a column vector of size $n = 3$. Similarly, an n vector is a row vector of n numbers as

$$a = \begin{bmatrix} a_1 & a_2 & \cdots & a_n \end{bmatrix}. \tag{2.2}$$

For example, $a = \begin{bmatrix} 1 & 2 & -3 \end{bmatrix}$ is a row vector of size n=3. We denote $\mathbb{R}$ as the set of real numbers and $\mathbb{R}^n$ is the set of column or row n-vectors with real components. We can say $\mathbb{R}^n$ as n-dimensional real vector space. We can denote the vectors by lowercase letters such as a, b, c, etc. The components of $a \in \mathbb{R}^n$ are denoted as a_1, a_2, $\ldots$, a_n.

The transpose (denoted as T) of a given column vector (2.1) is a row vector (2.2). Therefore, we can write

$$\begin{bmatrix} a_1 \\ a_2 \\ \vdots \\ a_n \end{bmatrix}^T = \begin{bmatrix} a_1 & a_2 & \cdots & a_n \end{bmatrix}.$$

The transpose of a row vector (2.2) is a column vector (2.1).

$$\begin{bmatrix} a_1 & a_2 & \ldots & a_n \end{bmatrix}^T = \begin{bmatrix} a_1 \\ a_2 \\ \vdots \\ a_n \end{bmatrix},$$

that is

$$a^T = \begin{bmatrix} a_1 \\ a_2 \\ \vdots \\ a_n \end{bmatrix}.$$

Note that the set of all row vectors forms a vector space called "row space", similarly the set of all column vectors forms a vector space called "column space".

A vector space V is a collection of vectors, which is closed under the operations of addition of two vectors $a, b \in V$, and multiplication by a scalar, $\alpha \in \mathbb{R}$, then the following properties hold:

1. Commutativity of vector addition: for vectors $a, b \in V$

$$a + b = b + a.$$

2. Associativity of vector addition: for vectors $a, b, c \in V$

$$a + (b + c) = (a + b) + c.$$

3. Existence of zero vector: for vector $a \in V$, we have

$$a + 0 = 0 + a = a.$$

4. Distributivity: for vectors $a, b \in V$ and scalars $\alpha, \beta \in \mathbb{R}$, we have

$$\alpha(a + b) = \alpha a + \alpha b,$$
$$(\alpha + \beta)a = \alpha a + \beta a.$$

5. Associativity of multiplication: for vector $a \in V$ and scalars $\alpha, \beta \in \mathbb{R}$, we have

$$\alpha(\beta a) = (\alpha \beta)a.$$

6. Unitarity: for vector $a \in V$, we have

$$1a = a.$$

7. The scalar 0 satisfies: for vector $a \in V$, we have

$$0a = 0.$$

8. Any scalar $\alpha \in \mathbb{R}$ satisfies:

$$\alpha 0 = 0.$$

9. Existence of negatives: for $a \in V$, we have

$$(-1)a = -a.$$

Two vectors $a = \begin{bmatrix} a_1 & a_2 & \ldots & a_n \end{bmatrix}^T$ and $b = \begin{bmatrix} b_1 & b_2 & \ldots & b_n \end{bmatrix}^T$ are equal if and only if $a_i = b_i$, for all $i = 1, 2, \ldots, n$.
We can add two vectors a and b as

$$a + b = \begin{bmatrix} a_1 + b_1 & a_2 + b_2 & \ldots & a_n + b_n \end{bmatrix}^T.$$

We can subtract two vectors a and b as

$$a - b = \begin{bmatrix} a_1 - b_1 & a_2 - b_2 & \ldots & a_n - b_n \end{bmatrix}^T.$$

The vector $0 - b$ is denoted as $-b$.

Suppose that $x = \begin{bmatrix} x_1, x_2, \ldots, x_n \end{bmatrix}^T$ is a solution to $a + x = b$. Then,

$$a_1 + x_1 = b_1,$$
$$a_2 + x_2 = b_2,$$
$$\vdots$$
$$a_n + x_n = b_n,$$

and thus

$$x = b - a.$$

We can say that vector $b - a$ is the unique solution of the vector equation $a + x = b$.

We define an operation of multiplication of a vector $a \in \mathbb{R}^n$ by a real scalar $\alpha \in \mathbb{R}$ as

$$\alpha a = \begin{bmatrix} \alpha a_1 & \alpha a_2 & \cdots & \alpha a_n \end{bmatrix}.$$

Note that $\alpha a = 0$ if and only if $\alpha = 0$ or $a = 0$. To see this, observe that $\alpha a = 0$ is equivalent to $\alpha a_1 = \alpha a_2 = \cdots = \alpha a_n = 0$. If $\alpha = 0$ or $a = 0$, then $\alpha a = 0$. If $a \neq 0$, then at least one of its components $a_k \neq 0$. For this component, $\alpha a_k = 0$, and hence we must have $\alpha = 0$. Similar arguments can be applied to the case when $\alpha \neq 0$.

Definition 2.1 (Linearly Independent). A set of vectors $S = \{a_1, a_2, \ldots, a_k\}$ is said to be linearly independent if the equality $\alpha_1 a_1 + \alpha_2 a_2 + \cdots + \alpha_k a_k = 0$ implies that all coefficients $\alpha_i \in \mathbb{R}$, where $i = 1, 2, \ldots, k$ are equal to zero.

Example 2.1. Prove that the vectors $a_1 = \begin{bmatrix} 1 \\ 0 \\ 1 \end{bmatrix}$, $a_2 = \begin{bmatrix} 0 \\ 1 \\ -1 \end{bmatrix}$, $a_3 = \begin{bmatrix} 0 \\ 0 \\ -1 \end{bmatrix}$ are linearly independent.

We apply definition of linear independent. We must show that the linear combination of vectors a_1, a_2 and a_3 are equal to zero in which all the coefficients α_1, α_2, and α_3 should be zero. Therefore, we can write as

$$\alpha_1 \begin{bmatrix} 1 \\ 0 \\ 1 \end{bmatrix} + \alpha_2 \begin{bmatrix} 0 \\ 1 \\ -1 \end{bmatrix} + \alpha_3 \begin{bmatrix} 0 \\ 0 \\ -1 \end{bmatrix} = \begin{bmatrix} 0 \\ 0 \\ 0 \end{bmatrix}.$$

Equating the corresponding coordinates of the vectors on the left and right side, we get the following system of linear equations:

$$\begin{aligned} \alpha_1 \phantom{{}-{}\alpha_2 - \alpha_3} &= 0, \\ \alpha_2 \phantom{{}- \alpha_3} &= 0, \\ \alpha_1 - \alpha_2 - \alpha_3 &= 0. \end{aligned}$$

Solving the above equations, we get $\alpha_1 = \alpha_2 = \alpha_3 = 0$. Thus, vectors a_1, a_2, and a_3 are linearly independent.

Definition 2.2 (Linearly Dependent). A set of the vectors $S = \{a_1, a_2, \ldots, a_k\}$ is said to be linearly dependent if there exists coefficients $\alpha_i \in \mathbb{R}$, where $i = 1, 2, \ldots, k$ not all of which are zero such that $\alpha_1 a_1 + \alpha_2 a_2 + \cdots + \alpha_k a_k = 0$.

Example 2.2. Show that the vectors $a_1 = \begin{bmatrix} 1 \\ 2 \\ 1 \end{bmatrix}$, $a_2 = \begin{bmatrix} 1 \\ -1 \\ 2 \end{bmatrix}$, and

$a_3 = \begin{bmatrix} 3 \\ 3 \\ 4 \end{bmatrix}$ are linearly dependent.

The vectors a_1, a_2, a_3 are linearly dependent because $2a_1 + a_2 - a_3 = 0$, where $\alpha_i \neq 0$, i.e., $\alpha_1 = 2, \alpha_2 = 1$, and $\alpha_3 = -1$.

Theorem 2.1. *A set of vectors $\{a_1, a_2, \ldots, a_k\}$ is linearly dependent if and only if one of the vectors a_i from the set is a linear combination of the remaining vectors.*

Proof. Using definition (2.2), since $\{a_1, a_2, \ldots, a_n\}$ is linearly dependent, there exists coefficients $\alpha_i \in \mathbb{R}$, not all zero such that

$$\alpha_1 a_1 + \cdots + \alpha_i a_i + \cdots + \alpha_k a_k = 0. \tag{2.3}$$

Suppose $\alpha_i \neq 0$ for some i, that is

$$a_i = -\frac{\alpha_1}{\alpha_i} a_1 - \frac{\alpha_2}{\alpha_i} a_2 - \cdots - \frac{\alpha_{i-1}}{\alpha_i} a_{i-1} - \frac{\alpha_{i+1}}{\alpha_i} a_{i+1} - \cdots - \frac{\alpha_k}{\alpha_i} a_k.$$

Conversely, for some i, a_i can be expressed as a linear combination of other vectors. That is,

$$a_i = \alpha_1 a_1 + \cdots + \alpha_{i-1} a_{i-1} + \alpha_{i+1} a_{i+1} + \cdots + \alpha_k a_k,$$

then we can write

$$\alpha_1 a_1 + \cdots + (-1) a_i + \alpha_{i+1} a_{i+1} + \cdots + \alpha_k a_k = 0.$$

Since $\alpha_i = -1 \neq 0$, thus, the set of vectors $\{a_1, a_2, \ldots, a_n\}$ is linearly dependent. $\qquad \square$

2.2   Matrix

A matrix is a rectangular array of numbers, commonly denoted by uppercase bold letters (e.g., A,B, etc.). A matrix with m rows and n columns is called an $m \times n$ matrix, and we write

$$A = \begin{bmatrix} a_{11} & a_{12} & \cdots & a_{1n} \\ a_{21} & a_{22} & \cdots & a_{2n} \\ \vdots & \vdots & \ddots & \vdots \\ a_{m1} & a_{m2} & \cdots & a_{mn} \end{bmatrix}.$$

The real number, a_{ij}, located in the ith row and j^{th} column is called the $(i,j)^{\text{th}}$ entry. We can think of A in terms of its n columns, each of which is a column vector in $\mathbb{R}^m$. Alternatively, we can think of A in terms of its m rows, each of which is a row n-vector. The transpose of matrix A, denoted as A^T, is the $n \times m$ matrix.

$$A^T = \begin{bmatrix} a_{11} & a_{21} & \cdots & a_{m1} \\ a_{12} & a_{22} & \cdots & a_{m2} \\ \vdots & \vdots & \ddots & \vdots \\ a_{1n} & a_{2n} & \cdots & a_{mn} \end{bmatrix}.$$

We see that columns of A are the rows of A^T and vice versa.
Note that the symbol $\mathbb{R}^{m \times n}$ denotes the set of $m \times n$ matrices whose entries are real numbers. We treat column vectors in $\mathbb{R}^n$ as elements of $\mathbb{R}^{n \times 1}$. Similarly, we treat row n-vectors as elements of $\mathbb{R}^{1 \times n}$.

2.3   Linear Equations

Consider m linear equations in n unknowns namely of $x_1, x_2, \ldots, x_n$ as:

$$a_{11}x_1 + a_{12}x_2 + \cdots + a_{1n}x_n = b_1,$$
$$a_{21}x_1 + a_{22}x_2 + \cdots + a_{2n}x_n = b_2,$$
$$\vdots$$
$$a_{m1}x_1 + a_{m2}x_2 + \cdots + a_{mn}x_n = b_m.$$

Equivalently,

$$Ax=b.$$

Associated with this system of equations is the matrix:

$$A = [a_1, a_2, \ldots, a_n].$$

Consider the $m \times n$ matrix

$$A = \begin{bmatrix} a_{11} & a_{21} & \cdots & a_{m1} \\ a_{12} & a_{22} & \cdots & a_{m2} \\ \vdots & \vdots & \ddots & \vdots \\ a_{1n} & a_{2n} & \cdots & a_{mn} \end{bmatrix}.$$

We can apply elementary row operations in the matrix A to get the matrix in reduced form.

An elementary row operation on the given matrix A is an algebraic manipulation of the matrix that corresponds to one of the following:

1. Interchanging any two rows such as the pth and the uth rows of the matrix A;

2. Multiplying one of its rows such as the pth row by a real number α where $\alpha \neq 0$;

3. Adding one of its rows such as the uth row to the β times p^{th} row.

Rank of Matrix

The number of nonzero rows in the row reduced form of a matrix A is called a rank of the matrix A, denoted as $\rho(A)$. It is read as 'rho of A'. Note that if the matrix A is of order $m \times n$ and $\rho(A) = m$, then A is said to be of full rank.

Example 2.3. Find the rank of the matrix $A = \begin{bmatrix} 2 & 2 & 2 & -2 \\ 1 & 2 & 3 & 4 \\ 3 & 4 & 5 & 2 \end{bmatrix}$.

$$\begin{bmatrix} 2 & 2 & 2 & -2 \\ 1 & 2 & 3 & 4 \\ 3 & 4 & 5 & 2 \end{bmatrix}$$

$R_1 \rightarrow \frac{1}{2}R_1$

$$\begin{bmatrix} 1 & 1 & 1 & -1 \\ 1 & 2 & 3 & 4 \\ 3 & 4 & 5 & 2 \end{bmatrix}$$

$R_2 \rightarrow R_2 - R_1$

$$\begin{bmatrix} 1 & 1 & 1 & -1 \\ 0 & 1 & 2 & 5 \\ 3 & 4 & 5 & 2 \end{bmatrix}$$

$R_3 \rightarrow R_3 - 3R_1$

$$\begin{bmatrix} 1 & 1 & 1 & -1 \\ 0 & 1 & 2 & 5 \\ 0 & 1 & 2 & 5 \end{bmatrix}$$

$R_3 \rightarrow R_3 - R_2$

$$\begin{bmatrix} 1 & 1 & 1 & -1 \\ 0 & 1 & 2 & 5 \\ 0 & 0 & 0 & 0 \end{bmatrix}.$$

Therefore, $\rho(A)$=Number of nonzero rows=2.

Example 2.4. Find the rank of the matrix.

$$\begin{bmatrix} 0 & 1 & -3 & -1 \\ 1 & 0 & 1 & 1 \\ 3 & 1 & 0 & 2 \\ 1 & 1 & -2 & 0 \end{bmatrix}.$$

Applying elementary row operations,

$$\begin{bmatrix} 0 & 1 & -3 & -1 \\ 1 & 0 & 1 & 1 \\ 3 & 1 & 0 & 2 \\ 1 & 1 & -2 & 0 \end{bmatrix}$$

$R_2 \leftrightarrow R_1$

$$\begin{bmatrix} 1 & 0 & 1 & 1 \\ 0 & 1 & -3 & -1 \\ 3 & 1 & 0 & 2 \\ 1 & 1 & -2 & 0 \end{bmatrix}$$

$R_3 \rightarrow R_3 - 3R_1$

$$\begin{bmatrix} 1 & 0 & 1 & 1 \\ 0 & 1 & -3 & -1 \\ 0 & 1 & -3 & -1 \\ 1 & 1 & -2 & 0 \end{bmatrix}$$

$R_4 \rightarrow R_4 - R_1$

$$\begin{bmatrix} 1 & 0 & 1 & 1 \\ 0 & 1 & -3 & -1 \\ 0 & 1 & -3 & -1 \\ 0 & 1 & -3 & -1 \end{bmatrix}$$

$R_3 \rightarrow R_3 - R_2$

$$\begin{bmatrix} 1 & 0 & 1 & 1 \\ 0 & 1 & -3 & -1 \\ 0 & 0 & 0 & 0 \\ 0 & 1 & -3 & -1 \end{bmatrix}$$

$R_4 \rightarrow R_4 - R_2$

$$
\begin{bmatrix}
1 & 0 & 1 & 1 \\
0 & 1 & -3 & -1 \\
0 & 0 & 0 & 0 \\
0 & 0 & 0 & 0
\end{bmatrix}.
$$

Therefore, $\rho(A)$=Number of nonzero rows=2.

The system of linear equations is said to be

1. Consistent if $\rho(A) = \rho(A|b)$, then

   (a) The system has a unique solution if $\rho(A) = \rho(A|b)$ =Number of variables.

   (b) The system has infinitely many solutions if $\rho(A) = \rho(A|b)<$Number of variables.

2. Inconsistent if $\rho(A) \neq \rho(A|b)$, then the system has no solution.

Example 2.5. Solve the following system of equations.

$$
\begin{aligned}
2x + 6y &= -11, \\
6x + 20y - 6z &= - 3, \\
6y - 18z &= - 1.
\end{aligned}
$$

We can write system of linear equations as an augmented matrix:

$$
\left[\begin{array}{ccc|c}
2 & 6 & 0 & -11 \\
6 & 20 & -6 & -3 \\
0 & 6 & -18 & -1
\end{array}\right]
$$

We proceed with elementary row operations.

$R_2 \rightarrow R_2 - 3R_1$

$$
\left[\begin{array}{ccc|c}
2 & 6 & 0 & -11 \\
0 & 2 & -6 & 30 \\
0 & 6 & -18 & -1
\end{array}\right]
$$

$R_3 \rightarrow R_3 - 3R_2$

$$\begin{bmatrix} 2 & 6 & 0 & -11 \\ 0 & 2 & -6 & 30 \\ 0 & 0 & 0 & -91 \end{bmatrix}.$$

We see that

$$\rho(A) = 2,$$
$$\rho(A|b) = 3.$$

That is,

$$\rho(A) \neq \rho(A|b).$$

Thus, the system is inconsistent and it has no solution.

2.4 Matrix Inversion

We apply the method of row reduction to find the inverse of a nonsingular matrix. If A be an $n \times n$ nonsingular matrix, then A^{-1} exists.

Suppose that we have an equation

$$Ax = b, \tag{2.4}$$

where $b \neq 0$. To solve this, we can proceed as follows:

$$A^{-1}(Ax) = A^{-1}b,$$
$$(A^{-1}.A)x = A^{-1}b.$$

Since

$$A^{-1}A = I,$$

then

$$Ix = A^{-1}b.$$

Since

$$Ix = x,$$

therefore

$$x = A^{-1}b. \tag{2.5}$$

Thus, solving (2.4) is just an equivalent to finding (2.5). But, the solution can be done by the process of row reduction. The method of row reduction will also be adaptable to find $A^{-1}b$ and consequently A^{-1}. We can also write (2.4) as

$$Ax = Ib. \tag{2.6}$$

Premultiplying (2.6) by A^{-1},

$$A^{-1}Ax = A^{-1}Ib.$$

We get

$$Ix = A^{-1}b. \tag{2.7}$$

In this process, we always work from A and arrive at I. From (2.6) and (2.7), it is observed that if the same row reduction is applied to identity matrix, I, then we end up with matrix A^{-1}.

Example 2.6. Find the inverse of the matrix $A = \begin{bmatrix} 1 & 1 & 0 \\ 1 & -1 & 1 \\ 1 & -1 & 2 \end{bmatrix}$.

$$\begin{bmatrix} 1 & 1 & 0 & 1 & 0 & 0 \\ 1 & -1 & 1 & 0 & 1 & 0 \\ 1 & -1 & 2 & 0 & 0 & 1 \end{bmatrix}$$

$R_2 \to R_2 - R_1$

$$\begin{bmatrix} 1 & 1 & 0 & 1 & 0 & 0 \\ 0 & -2 & 1 & -1 & 1 & 0 \\ 1 & -1 & 2 & 0 & 0 & 1 \end{bmatrix}$$

$R_3 \to R_3 - R_1$

$$\begin{bmatrix} 1 & 1 & 0 & 1 & 0 & 0 \\ 0 & -2 & 1 & -1 & 1 & 0 \\ 0 & -2 & 2 & -1 & 0 & 1 \end{bmatrix}$$

$R_2 \to -\frac{1}{2}R_2$

$$\begin{bmatrix} 1 & 1 & 0 & 1 & 0 & 0 \\ 0 & 1 & -1/2 & 1/2 & -1/2 & 0 \\ 0 & -2 & 2 & -1 & 0 & 1 \end{bmatrix}$$

$R_3 \to \frac{1}{2}R_3$

$$\begin{bmatrix} 1 & 1 & 0 & 1 & 0 & 0 \\ 0 & 1 & -1/2 & 1/2 & -1/2 & 0 \\ 0 & -1 & 1 & -1/2 & 0 & 1/2 \end{bmatrix}$$

$R_1 \to R_1 - R_2$

$$\begin{bmatrix} 1 & 0 & 1/2 & 1/2 & 1/2 & 0 \\ 0 & 1 & -1/2 & 1/2 & -1/2 & 0 \\ 0 & -1 & 1 & -1/2 & 0 & 1/2 \end{bmatrix}$$

$R_3 \to R_3 + R_2$

$$\begin{bmatrix} 1 & 0 & 1/2 & 1/2 & 1/2 & 0 \\ 0 & 1 & -1/2 & 1/2 & -1/2 & 0 \\ 0 & 0 & 1/2 & 0 & -1/2 & 1/2 \end{bmatrix}$$

$R_3 \to 2R_3$

$$\begin{bmatrix} 1 & 0 & 1/2 & 1/2 & 1/2 & 0 \\ 0 & 1 & -1/2 & 1/2 & -1/2 & 0 \\ 0 & 0 & 1 & 0 & -1 & 1 \end{bmatrix}$$

$R_1 \to R_1 - \frac{1}{2}R_3$

$$\begin{bmatrix} 1 & 0 & 0 & 1/2 & 1 & -1/2 \\ 0 & 1 & -1/2 & 1/2 & -1/2 & 0 \\ 0 & 0 & 1 & 0 & -1 & 1 \end{bmatrix}$$

$R_2 \rightarrow R_2 + \frac{1}{2}R_3$

$$\begin{bmatrix} 1 & 0 & 0 & 1/2 & 1 & -1/2 \\ 0 & 1 & 0 & 1/2 & -1 & 1/2 \\ 0 & 0 & 1 & 0 & -1 & 1 \end{bmatrix}.$$

Therefore,

$$A^{-1} = \begin{bmatrix} 1/2 & 1 & -1/2 \\ 1/2 & -1 & 1/2 \\ 0 & -1 & 1 \end{bmatrix}.$$

2.5   Exercises

Exercise 2.1. Determine if each of the following sets of vectors is linearly independent or linearly dependent.

(a) $V=\left\{\begin{bmatrix}1 & 0 & 1\end{bmatrix}, \begin{bmatrix}1 & 2 & 1\end{bmatrix}, \begin{bmatrix}2 & 2 & 2\end{bmatrix}\right\}$

(b) $V=\left\{\begin{bmatrix}2 & 1 & 0\end{bmatrix}, \begin{bmatrix}1 & 2 & 0\end{bmatrix}, \begin{bmatrix}3 & 3 & 1\end{bmatrix}\right\}$

(c) $V=\left\{\begin{bmatrix}2 & 1\end{bmatrix}, \begin{bmatrix}1 & 2\end{bmatrix}\right\}$

Exercise 2.2. Find the rank of the following matrices.

(a)

$$A = \begin{bmatrix} 2 & 3 & -1 & -1 \\ 1 & -1 & -2 & -4 \\ 3 & 1 & 3 & -2 \\ 6 & 3 & 0 & -7 \end{bmatrix}$$

(b)

$$A = \begin{bmatrix} 6 & 1 & 3 & 8 \\ 16 & 4 & 12 & 15 \\ 5 & 3 & 3 & 4 \\ 4 & 2 & 6 & -1 \end{bmatrix}$$

Exercise 2.3. Find the inverse of the following matrices.

(a)

$$A = \begin{bmatrix} 5 & 6 \\ 3 & -2 \end{bmatrix}$$

(b)

$$A = \begin{bmatrix} 1 & 2 & -4 \\ -1 & -1 & 5 \\ 2 & 7 & -3 \end{bmatrix}$$

Exercise 2.4. Find the inverse of the following matrices.

(a)

$$A = \begin{bmatrix} 1 & 1 & 1 & 1 \\ 0 & 1 & 1 & 1 \\ 0 & 0 & 1 & 1 \\ 0 & 0 & 0 & 1 \end{bmatrix}$$

(b)

$$A = \begin{bmatrix} 1 & 2 & 1 & 0 \\ 0 & 1 & -1 & 1 \\ 1 & 3 & 1 & -2 \\ 1 & 4 & -2 & 4 \end{bmatrix}$$

Exercise 2.5. Solve the following system of equations.

(a)

$$\begin{aligned} x + y + 2z &= 4, \\ 2x + 3y + 6z &= 10, \\ 3x + 6y + 10z &= 14. \end{aligned}$$

(b)

$$\begin{aligned} 2x - y - 4z &= 2, \\ 4x - 2y - 6z &= 5, \\ 6x - 3y - 8z &= 8. \end{aligned}$$

Chapter 3

MATLAB

3.1 Introduction

MATLAB is a standard tool which has been included in introductory and advanced courses in applied mathematics, engineering, science and economics in many universities around the world. In industry, it is a tool of research, development, and analysis. MATLAB (short for MATrix LABoratory) is a mathematical and graphical software package with numerical, graphical, and programming capabilities developed by the MathWorks, Inc., Natick, Massachusetts, USA. In 1984, the first version appeared.

MATLAB is helpful to solve complicated problems. All major functions can directly be used as the input. MATLAB uses an interpreter to understand what we type and what type of output may come. Users sometimes get suggestions to make error-free statements. Therefore, we need to be careful. We begin by studying the basic feature in MATLAB.

3.2 Basic Feature

To start MATLAB in Microsoft Windows, double-click on the MATLAB icon on the Windows desktop. MATLAB can also be started by selecting MATLAB from the Start menu. Once MATLAB is launched, a MATLAB Command Window appears on the screen.

Command Window
The Command Window is used to enter several variables, evalu-

ate MATLAB commands, and run M-files or functions.

Edit Window

To open the Edit Window, go to File→New→Script. This window allows us to type and save a series of commands without executing them. We can also open the Edit Window by typing Edit at the Command prompt or by selecting the New Script button on the toolbar.

MuPAD

MuPAD opens a new blank MuPAD notebook and returns an object representing the notebook. A MuPAD notebook is an easy to use environment for performing computations symbolically using the MuPAD language and documenting the results.
For example, we write in the Command Window:

```
>> Mupad
```
It will open a new blank MuPAD notebook.

Help Browser

The Help Browser is used to view pre-defined documentation for all MATLAB products. It helps to know about any command that we want.
For example, to know about `for` loop, we write in the Command Window as follows:

```
>> help for
```

3.3   Basic Operations in MATLAB

In the Command Window, we see: >>. This notation >> is called the prompt.
In the Command Window, MATLAB can be used interactively. It

means that MATLAB command or expression can be entered, and MATLAB immediately responds with the result.

For example, this is the way it would appear in the Command Window:

```
>> mynum = 10
mynum =
        10
```

MATLAB uses a default variable named, if an expression is typed at the prompt and it is not assigned to a variable. For example, the result of the expression $9 + 3$ is stored in the variable ans.

```
>> 9+3
ans =
        12
```

Variables and Assignments

In MATLAB, we use the equal sign to assign values to a variable. For example:

```
>> u = 9

u =
     9
```

Henceforth, MATLAB always takes the value of the variable u as 9. For example:

```
>> u*2 - 2*u + u
ans =
        9
```

Note that MATLAB never forgets used variables unless instructed to do so. We can check the current value of a variable by simply typing its name.

Example 3.1. $a = 4$, $b = 6$, $c = 3$. Find $a \times (b + c)$, $a \times (b + c)$, $\frac{a}{b} + c$, and $\frac{a}{b+c}$.

In the Command Window,

```
>> format rat
>> a = 4; b = 6; c = 3;
>> a*(b+c)
>> a*b+c
>> a/b+c
>> a/(b+c)
```

In this example, we get the answers $36, 27, {}^{11}/_3, {}^4/_9$. This gives some idea that MATLAB performs those calculations first which are in brackets. The command `format rat` has been used to force the results to be shown as rationals, the final command format reverts to the default. We have used a semicolon at the end of MATLAB assignment statements to suppress echoing of assigned values in the Command Window. This greatly speeds program execution.

The transpose operator swaps the row and columns of any array that it is applied to.
For example:

```
>> f = [1:4]'
f =
     1
     2
     3
     4
```

Vectors and Matrices

A vector is a list of elements. Elements should be separated by comma or space for row vector and semicolon for column vector. For example:

```
>> B = [5 6 7]
B =
     5        6        7
```

```
>> b= [23; 34; 12]
```

b =
```
    23
    34
    12
```

A matrix is a rectangular array of numbers. Row and column vectors are also examples of matrices.
For example:
```
>> A = [1 3 5 6; 7 -3 1 8; 5 1 -1 9]
```

A =
1	3	5	6
7	-3	1	8
5	1	-1	9

MATLAB has many built-in functions. The built-in function **zeros** can be used to create an all-zero array of any desired size.
For example:

```
>> a = zeros(2)
```

a =
```
  0   0
  0   0
```

```
>> b = zeros(2,3)
```

b =
```
  0     0     0
  0     0     0
```

eye function can be used to generate arrays containing identity matrices, in which all on-diagonal elements are one, while all off-diagonal elements are zero.
For example:

```
>> eye(3)
```

ans =
1	0	0
0	1	0
0	0	1

size() Function

`size()` function returns two values specifying the number of rows and columns of any matrix.
For example:

```
>> A=[1 2 3 7 4; 4 5 8 4 -3; 0 5 9 0 -3; 2 1 7 3 0 ];
```

```
>> [m n]=size(A)
```

m =

    4

n =

    5

end Function

MATLAB provides a special function named **end** that is very useful for creating array subscripts. The **end** function returns the highest value taken by the subscript in vector or matrix.
For example:

$$>> B = \begin{bmatrix} 8 & 7 & 6 & 5 & 4 & 3 & 2 & 1 \end{bmatrix};$$

```
>> B(5 : end)
```

ans =

      4   3   2   1

We can write 3×4 matrix in MATLAB as follows:

$$>> A = \begin{bmatrix} 1 & 2 & 3 & 4; & 5 & 6 & 7 & 8; & 9 & 10 & 11 & 12 \end{bmatrix}$$

A =

1	2	3	4
5	6	7	8
9	10	11	12

```
>> A(2: end, 2:end)
```

ans =

6	7	8
10	11	12

abs() Function

`abs()` function calculates the absolute value of any variable. For example:

```
>> x= abs(-4)
```

```
x =
    4
```

Comments

Comments are an integral part of any programming language. Comments help to identify program purpose and explain the work of particular statements in a program. Comments also allow others to understand the code.

To comment out multiple lines of code, we can use the block comment operators, %{ and %} :

```
%{
Introduction to LINEAR PROGRAMMING with MATLAB
Chapter I– By S K Mishra
%}
```

We can also create quick, one-line comments with operator %. The next line of code demonstrates this.

```
% By S K Mishra
```

MATLAB SCRIPTS

A Script is nothing but, a computer program written in the language of MATLAB. It is stored in an M-file. It is saved with extension `.m`. We can display the contents of the script in the Command Window using the `type` command followed by file name without `.m` extension. Interpreter is a computer program which executes the statements of script step by step. The script can be executed, or run, by simply entering the name of the file (without the .m extension) in the command window. MATLAB ignores the comment lines and does not execute when we run the M-file. For example:

Code 3.1: radius.m.

```
%find radius of circle
radius=5;
area=pi*(radius)^2;
```

Note that **type** command is helpful to see the contents of the script in the Command Window. For eample:

```
>> type radius
```

Output:

```
radius=5;
area=pi*(radius)^2
```

To run the script, the name of the file is entered at the prompt (again, without the .m) or press **F5**.

```
>> radius
radius =
        5
area =
      78.5398
```

INPUT/OUTPUT

Statements that accomplish to print the output or take the input from users are called Input/Output statements. Input statements read in values from the standard input device that is the keyboard.

In the Command Window,

```
>> side = input('Enter the side of square')
```

Output:

```
Enter the side of square:8
side =
      8
```

If we want character or string input, then 's' must be added as a second argument to the input function:

In the Command Window,

```
>> letter = input('Enter a char','s');
```
Output:

Enter a char: g
letter=
      g

The simplest output function in MATLAB is `disp`, which is used to display the result of an expression.
For example:

In the Command Window,

```
>> disp('Hello')
```

Output:

    Hello

In the Command Window,

```
>> disp(4^3)
```

Output:

    64

`fprintf()` displays the values of several variables in a specified format. For eample:

In the Command Window,

```
>>a=[1  2  3  4  5];
```

```
>> fprintf('%d',a)
```

Output:

12345>>

We need to know different types of specifiers which are shown in Table 3.1.

TABLE 3.1: Types of Specifiers

Specifiers	Type
%c	Character
%s	String
%d	Decimal integer number
%f	Floating point number

Variable Precision

MATLAB uses floating-point arithmetic for its calculations. Using the Symbolic Math Toolbox, we can also do exact arithmetic with symbolic expressions. For example:

In the Command Window,

```
>>x = cot(pi/2)
```

Output:

```
x =
    6.1232e-17
```

The value of x is in floating-point format that is 6.1232×10^{-17}. However, we know that $\cot(\pi/2)$ is equal to 0. This inaccuracy is due to the fact that typing pi in MATLAB gives an approximation to π accurate to about 15 digits, not its exact value. To compute an exact value, we must type `sym(pi/2)`.
For example:

In the Command Window,

```
>> x = cot(sym(pi/2))
```

Output:

```
x =
    0
```

We wanted this value.

format

`format Command` style changes the output display format in the Command Window to the format specified by style. For example:

In the Command Window,

```
>> format long
>> pi
```

Output:

```
ans =
      3.141592653589793
```

See Table 3.2 for several numeric display of formats.

TABLE 3.2: Numeric Display of Formats

Type	Result	Example
format short	5 digits	3.1416
format long	15 digits	3.141592653589793
format short g	Best of fixed or floating point, with 5 digits	3.1416
format long g	15 digits	3.14159265358979
format short eng	Engineering format	3.1416e+000
format long eng	Engineering format with 16 significant digits and a power is a multiple of three	3.14159265358979e+000

3.4 Selection Statements and Loop Statements

If the expression is true, then the commands are executed, otherwise the program continues with the next command immediately beyond the end statement.

The simplest form of **if** statements is

$$
\begin{aligned}
&\text{if } <\text{condition}>\\
&\quad \text{statement1};\\
&\text{end}
\end{aligned}
$$

If the condition is true, the statement1 is executed, but if the condition is false, nothing happens. We use relational operators to create condition in **if** statement. Several relational operators have been shown in Table 3.3.

TABLE 3.3: Relational Operators

Operators	Meaning
$<$	Less than
$>$	Greater than
$<=$	Less than or equal
$>=$	Greater than or equal
$==$	Equivalent
$\sim=$	Not equal to

Suppose that an expression is $A < B$. When this condition is true, then block of statements will be executed. When this condition is false, then block of statements will not be executed.

The syntax of an **if...else** statement in MATLAB is

$$
\begin{aligned}
&\text{if } \text{condition}\\
&\quad \text{statementA};\\
&\text{else}\\
&\quad \text{statementB};\\
&\text{end}
\end{aligned}
$$

If the condition is true, then the **if** block of code will be executed, otherwise **else** block of code will be executed.

Example 3.2. Write MATLAB script to find whether a number is negative or not.

See MATLAB function given **neg.m** in Code 3.2.

Code 3.2: neg.m

```
function  a=neg(a)
if  a<0
```

```
        disp ('Negative')
    else
        disp ('Positive')
    end
```

In the Command Window,

```
>> neg(-2)
```

Output:

```
negative
```

The **for** loop is a loop that executes a block of statements a specified number of times. The syntax of **for** loop has the form:

```
for index = values
        statement1;
        statement2;
            . . .
    end
```

The loop begins with the **for** statement and ends with the **end** statement.

Example 3.3. Write MATLAB script to compute and display 10!.

We have written MATLAB script in the following Code 3.3.

Code 3.3: factorial.m

```
function f=factorial(a)
f=1;
for n=2:a
        f=f*n;
end
```

Example 3.4. Create a script file to display numbers from 1 to 10.

See Code 3.4.

Code 3.4: increment.m

```
for n=1:10
        disp(n);
    end
```

Example 3.5. Write MATLAB function to display the most negative element in a given matrix

$$A = \begin{bmatrix} -1 & -4 & -3 \\ -5 & -2 & -6 \\ -7 & -9 & -8 \end{bmatrix}.$$

This MATLAB Code 3.5 is used to find the most negative element that is –9 in matrix A.

Code 3.5: mostnegative.m

```
function mn=mostnegative(A)
%input matrix A
%find most negative element in matrix A
[m,n]=size(A);
mn=0;
for I=1:m
    for J=1:n
        if A(I,J)<= mn
            mn=A(I,J);
        end
    end
end
return
```

In the Command Window:

```
>> A=[-1 -4 -3; -5 -2 -6; -7 -9 -8 ]
>> mn = mostnegative(A)
```
Output:

```
mn =
    -9
```

The **while** loop is a loop that executes a block of statements repeatedly as long as the expression is true. The syntax is

```
while expression
    statement1;
    statement2;
    . . .
end
```

For example:

In the Command Window,
```
>> k = 0;
while k<3
k = k+1
end
```
Output:

```
k=
  1
k=
  2
k=
  3
```

The **for** and **while** loops can be terminated using the **break** command.

Example 3.6. Write a MATLAB function to display even numbers in vector.

See Code 3.6.

Code 3.6: even.m

```
function E = even(B)
% input  :  vector B
% output:  vector E
% This function displays even number
[~,n]=size(B);
j=1;
c=1;
while j<=n
        if mod(B(1,j),2)==0
            E(1,c)=B(1,j);
            c=c+1;
        end
        j=j+1;
end
return
```

3.5   User-Defined Function

A function is a collection of sequential statements that accepts an input argument from the user and provides output to the program. Functions allow us to program efficiently. It avoids rewriting the computer code for calculations that are performed frequently. User-defined functions are stored as M-files.

See a very simple MATLAB function `poly.m` in the following Code 3.7 that calculates the value of a particular polynomial.

Code 3.7: poly.m

```
function output = poly(x)
% This function calculates the value of third
% order
output=x^3 +x+3;
return
```

Note that file name should be the same as that of function. Therefore, we save as `poly.m`

In the Command Window,

```
>> poly(3)
```

Output:

```
output =
        33
```

We have used the elementary row operations in Linear Equations of **Chapter 2** to find the rank of matrix and solve the system of linear equations. We can write MATLAB code to perform these elementary operations and use them to solve several problems. For Example:

We have developed MATLAB function `exchange.m` given in the following Code 3.8 to exchange the elements of pth and u^{th} rows in any matrix.

Code 3.8: exchangeop.m

```
function A=exchangeop(A,p,u)
% input  :  augmented  matrix  A,  row  p,u
% output:  augmented  matrix  A
[~,n]=size(A);
for  J=1:n
     t=A(p,J);
     A(p,J)=A(u,J);
     A(u,J)=t ;
end
return
```

We have written MATLAB function `identityop.m` in the following Code 3.9 to place the identity element at any position in the matrix.

Code 3.9: identityop.m

```
function A=identityop(A,p,e)
%input  :  Augmented  matrix  A,  pivot
%row  p,  pivotelement  e
%output : Augmented  matrix  A
[~,n]=size(A);
format  rat
for  J=1:n
     A(p,J)=sym(e*A(p,J));
end
return
```

We have written a simple MATLAB function `eliminationop.m` in the following Code 3.10 to eliminate all elements of any row using pivot element in the matrix.

Code 3.10: eliminationop.m

```
function  A=eliminationop(A,u,p,co)
%input :  augmented  matrix  A,  row  u,
%pivotrow  p,  coefficientvalue  co
%output :  augmented  matrix  A
[~,n]=size(A);
format  rat
for  J=1:n
     A(u,J)=sym(A(u,J)+(co*A(p,J)));
```

```
        end
        return
```

Example 3.7. Find rank of matrix of **Example 2.3** in MATLAB.

$$A = \begin{bmatrix} 2 & 2 & 2 & -2 \\ 1 & 2 & 3 & 4 \\ 3 & 4 & 5 & 2 \end{bmatrix}.$$

```
>> A = identityop(A,1,1/2)
```

$$\begin{bmatrix} 1 & 1 & 1 & -1 \\ 1 & 2 & 3 & 4 \\ 3 & 4 & 5 & 2 \end{bmatrix}$$

```
>> A = eliminationop(A,2,1,-1)
```

$$\begin{bmatrix} 1 & 1 & 1 & -1 \\ 0 & 1 & 2 & 5 \\ 3 & 4 & 5 & 2 \end{bmatrix}$$

```
>> A = eliminationop(A,3,1,-3)
```

$$\begin{bmatrix} 1 & 1 & 1 & -1 \\ 0 & 1 & 2 & 5 \\ 0 & 1 & 2 & 5 \end{bmatrix}$$

```
>> A = eliminationop(A,3,2,-1)
```

$$\begin{bmatrix} 1 & 1 & 1 & -1 \\ 0 & 1 & 2 & 5 \\ 0 & 0 & 0 & 0 \end{bmatrix}.$$

Therefore, $\rho(A)$=Number of nonzero rows=2.

3.6   MATLAB Functions Defined in This Book

We list below the user-defined MATLAB functions. These functions will be used to solve several linear programming problems in this book.

1. help A=exchangeop(A,p,u)

    input   : augmented matrix A, row p,u
    output   : augmented matrix A
    operation : exchange elements of row p with
                elements of row u in tableau

2. help A=identityop(A,p,e)

    input   : augmented matrix A, pivot row p,
                pivotelement e
    output   : augmented matrix A
    operation : multiply real number e to
                elements of row p

3. help A=eliminationop(A,u,p,co)

    input   : augmented matrix A, row u,
                pivotrow p, coefficientvalue co
    output   : augmented matrix A
    operation : add elements of u to elements p
                times co

4. help A=simplex(A,p,q)

    input   : augmented matrix A, pivot row p,
                pivot column q
    output   : augmented matrix A
    operation : perform elementary row operation
                to make zero entries at q except
                unit entry at A(p,q)

5. help [A,q]=pivotcolumn(A,v)

    input   : augmented matrix A, nonbasic
                variables v
    output   : pivot column q, augmented
                matrix A
    operation : find pivot column in A

6. help [A,p,e,B]=pivotrow (A,q,B)

    input   : augmented matrix A, pivot
                    column q, basis
                    matrix B
    output   : augmented matrix A, pivot row p,
                    pivot element e
                    basis matrix B
    operation : find pivot row in A

7. help A=updatelastrow (A,av)

    input   : augmented matrix A, artificial
                    variables av
    output   : augmented matrix A
    operation : adding elements of all
                    corresponding rows except
                    last row and subtract
                    same from last row in A

8. help [A,p,q,e,B]=dual (A,B)

    input   : augmented matrix A, basis
                    matrix B
    output   : augmented matrix A, pivot
                    row p, pivot column q, pivot
                    element e, basis matrix B
    operation : find p,q,e,B using dual
                    simplex algorithm

9. help [Binv,B,xB]=rsm (A,c,B,xB,Binv,v)

    input   : augmented matrix A, cost c,
                    basis matrix B, basic vector
                    xB, identity matrix Binv,
                    nonbasic variable v
    output   : basis matrix B, basic vector xB
    operation : find B, xB using revised simplex
                    algorithm

10. help [minTcost,b,c]=nwc(A,sup,dem)

 input : transportation matrix A, supply
       sup, demand dem
 output : minimum transportation cost
       minTcost, basic matrix b, cost
       matrix c
 operation: find minTcost using northwest
       corner method

11. help [minTcost,b,c]=leastcost(A,sup,dem)

 input : transportation matrix A, supply
       sup, demand dem
 output : minimum transportation cost
       minTcost, basic matrix b, cost
       matrix c
 operation: find minTcost using least cost
       method

12. help [minTcost,b,c]=vogel(A,sup,dem)

 input : transportation matrix A, supply
       sup, demand dem
 output : minimum transportation cost
       minTcost, basic matrix b, cost
       matrix c
 operation: find minTcost using vogel's
       approximation method

13. help [u,v,b,c]=multipliers2(b,A,c,i,j)

 input : basic matrix b, transportation
       matrix A, cost matrix c, row i,
       column j
 output : vector u, vector v
 operation: find u and v

14. help x=uvx3(b,u,v,A)
    input   : basic matrix b, vector u and v,
                 transportation matrix A
    output   : nonbasic matrix x
    operation: solve nonbasic cells

15. help [basic,row,col]=mostpositive4(A,x,c)
    input   : transportation matrix A,
                 nonbasic matrix x, cost
                 matrix c
    output   : element basic, position row and
                 column
    operation: find most positive element in
                 nonbasic matrix x

16. help [y,bout]=cycle5(c,row,col,b)
    input   :cost matrix c, position row, col,
                 basic matrix b
    output   :loop matrix y,bout
    operation:find loop

17. help [c,b,min]=basiccell6(c,y,b,row,col)

    input   : cost matrix c, basic matrix b
    output   : minimum value min
    operation: find basic cell

18. help [C,T]=hungarian(A)

    input   :matrix A
    output   :optimal assignment C, optimal
                 value T
    operation:solve the assignment problem
                 using the hungarian method

3.7   Exercises

Exercise 3.1. Suppose that u= 3 and v= 4. Evaluate the following expressions using MATLAB.

(a) $\frac{5u}{3v}$

(b) $\frac{3v^{-2}}{(u+v)^2}$

(c) $\frac{v^3}{(v-u)^2}$

Exercise 3.2. Assume that a, b, c, and d are defined as follows:

$$a = \begin{bmatrix} 3 & 0 \\ 2 & 1 \end{bmatrix}, \qquad b = \begin{bmatrix} -1 & 2 \\ 0 & 1 \end{bmatrix}, \qquad c = \begin{bmatrix} 3 \\ 2 \end{bmatrix}, \qquad d = 1.$$

What is the result of each of the following expressions in MAT-LAB?

(a) a+b

(b) b*c

(c) a .* b

(d) a .* d

Exercise 3.3. Answer the following questions for the following array.

$$A = \begin{bmatrix} 1 & 2 & 8 & 5 & 3 \\ 6 & 8 & 1 & 3 & 1 \\ 12 & 6 & 9 & 1 & 0 \\ 5 & 8 & 1 & 6 & 4 \end{bmatrix}$$

(a) What is the size of A?

(b) What is the value of A(1,4)?

(c) What is the size and value of A(:,1:2:5)?

(d) What is the size and value of A([1 3], end)?

Exercise 3.4. Check the following expressions in the Command Window.

(a) $x = 4 > 2$

(b) $x = 2 > 5$

(c) $x = 4 <= 3$

(d) $x = 1 < 1$

(e) $x = 2\ = 2$

(f) $x = 3 == 3$

(g) $x = 0 < 0.5 < 1$

Exercise 3.5. Write the MATLAB function to find the following sum:

$$1^2 + 2^2 + 3^2 + \cdots + 1000^2$$

Exercise 3.6. Create a matrix B equal to [-1/3, 0, 1/3, 2/3], and use each of the built-in format options to display the results.

(a) `format short` (which is the default)

(b) `format long`

(c) `format bank`

(d) `format short e`

(e) `format long e`

(f) `format short eng`

(g) `format long eng`

(h) `format short g`

(i) `format long g`

(j) `format +`

(k) `format rat`

Chapter 4

Introduction to Linear Programming

4.1 Introduction

The objective of a linear programming problem is to obtain an optimal solution. Linear programming problems deal with the problem of minimizing or maximizing a linear objective function in the presence of a system of linear inequalities. The linear objective function represents cost or profit. A large and complex problem can be formulated in the form of a linear programming problem, and users can solve such a large problem in a definite amount of time using the simplex method and computer.

In this part we study a graphical method for solving linear programming problems. This method is helpful to choose the best feasible point among the many possible feasible points. A point minimizing the objective function and satisfying the set of linear constraints is called a "feasible point".

4.2 Simple Examples of Linear Programs

A linear programming problem is concerned with solving a very special type of problem—one in which all relations among the variables are linear both in the constraints and the function to be optimized. We wish to solve the linear programming problem. A linear programming problem is an optimization problem which can be written in standard form as

$$\begin{aligned} &\text{minimize} &&c^T x, \\ &\text{subject to} &&Ax = b, \\ &\text{where} &&x \geq 0, c \in \mathbb{R}^n, b \in \mathbb{R}^m, A \in \mathbb{R}^{m \times n}. \end{aligned}$$

The vector inequality $x \geq 0$ means that each component of x is nonnegative. In the above problem, $c^T x$ is called as an objective function to be minimize and $Ax = b$ is called as a set constraints. This problem is the case of minimization. We can also write maximize in place of minimize and in place $Ax = b$, we can write the inequalities such as $Ax \geq b$ or $Ax \leq b$ in the above linear programming problem. These inequalities can also be rewritten into the standard form as shown above.

The purpose of this section is to formulate the linear programming problems and illustrate the applications of linear programming methods.

Example 4.1. *(The Diet Problem)* Assume that there are two products, cereal and milk, for breakfast and assume that a person must consume at least 60 units of iron and at least 70 units of protein to stay alive. Assume that one unit of cereal costs \$20 and contains 30 units of iron and 5 units of protein and one unit of milk costs \$10 and contains 17 units of iron and 9 units of protein. The goal is to find the cheapest diet which will satisfy the minimum daily requirement.

Let x_1 represents the number of units of cereal that the person consumes a day and x_2 represents the number of units of milk consumed.

For the diet to meet the minimum requirements, we must have

Iron Requirement   : $30x_1 + 17x_2 \geq 60$,
Protein Requirement : $5x_1 + \ 9x_2 \geq 70$, where $x_1, x_2 \geq 0$.

The cost of the diet is: $20x_1 + 10x_2$.
Hence, the diet problem is:

$$\begin{aligned} &\text{minimize} &&20x_1 + 10x_2 \\ &\text{subject to} &&30x_1 + 17x_2 \geq 60, \\ & &&5x_1 + \ 9x_2 \geq 70, \\ & &&x_1, \quad x_2 \geq 0. \end{aligned}$$

Example 4.2. A manufacturer produces two different products, say chairs and tables, using three machines M_1, M_2 and M_3. Each machine can be used for only a limited period of time. Production time for each product on each machine is given below in Table 4.1.

TABLE 4.1: Production Time

Machine	Production time (hrs/unit) Chair	Table	Available time
M_1	1	1	18
M_2	1	3	18
M_3	2	1	14
total	4 hrs	5 hrs	

The objective is to maximize the combined time of utilization of all three machines.

Let x_1 and x_2 denote the number of chairs and tables.

Constraints:

$$x_1 + x_2 \leq 18, \; x_1 + 3x_2 \leq 18, \; 2x_1 + x_2 \leq 14, \text{ where } x_1, x_2 \geq 0.$$

Objective: maximizing the combined production time of three machines, that is

$$4x_1 + 5x_2.$$

Therefore, the linear programming problem is

$$\begin{aligned}
\text{maximize} \quad & 4x_1 + 5x_2 \\
\text{subject to} \quad & x_1 + x_2 \leq 18, \\
& x_1 + 3x_2 \leq 18, \\
& 2x_1 + x_2 \leq 14, \\
& x_1, \quad x_2 \geq 0.
\end{aligned}$$

Example 4.3. A person requires $8, 15, 16$ units of chemicals A, B, C, respectively, for his garden. The liquid product contains $4, 5, 3$ units of A, B, C, respectively per jar. The dry product contains $3, 5, 6$ units of A, B, C respectively per packet. The person wants to spend a minimum possible amount on his garden, where it is given that the liquid product is available for \$40 per jar and the dry product is available for \$25 per packet.

Suppose that person purchases x_1 jars and x_2 packets of the products.

Constraints:

$4x_1 + 3x_2 \geq 8$, $5x_1 + 5x_2 \geq 15$, $3x_1 + 6x_2 \geq 16$, where $x_1, x_2 \geq 0$.

Objective: Minimizing cost, that is

$$40x_1 + 25x_2.$$

Therefore, the linear programming problem is

$$
\begin{array}{rrrl}
\text{minimize} & 40x_1 + & 25x_2 & \\
\text{subject to} & 4x_1 + & 3x_2 \geq & 8, \\
& 5x_1 + & 5x_2 \geq & 15, \\
& 3x_1 + & 6x_2 \geq & 16, \\
& x_1, & x_2 \geq & 0.
\end{array}
$$

Example 4.4. A hotel has the following requirements for waiters shown in Table 4.2. Waiters report to the hotel rooms at the beginning of each period and work for eight consecutive hours. The same waiter cannot work for more than two consecutive periods. The hotel wants to determine the minimum number of waiters, so that there may be sufficient waiters available for each period. Formulate this as a linear programming problem.

TABLE 4.2: Hotel Requirement

Period	Clock time (24 hours per day)	Minimum number of waiters required
1	7 A.M. - 11 A.M.	65
2	11 A.M. - 3 P.M.	75
3	3 P.M. - 7 P.M.	65
4	7 P.M. - 11 P.M.	55
5	11 P.M. - 3 A.M.	25
6	3 A.M. - 7 A.M.	35

Suppose that $x_1, x_2, x_3, x_4, x_5, x_6$ be the number of waiters reporting at the beginning of periods $1, 2, \ldots, 6$.

Objective: minimizing the number of waiters, that is

$$\text{minimize } x_1 + x_2 + x_3 + x_4 + x_5 + x_6.$$

Since the same waiter cannot work for more than two consecutive periods, x_1 waiters work for the period of 1 and 2, x_2 waiters work for the period of 2 and 3 etc. But, for the period 1, the minimum number of waiters required is 65. Similarly, we can write the minimum number of waiters required for periods 2,3,4,5 and 6.

Constraint: $x_6 + x_1 \geq 65$.

Similarly, other constraints are $x_1 + x_2 \geq 75$, $x_2 + x_3 \geq 65$, $x_3 + x_4 \geq 55$, $x_4 + x_5 \geq 25$, $x_5 + x_6 \geq 35$.

Therefore, the linear programming problem is

$$
\begin{aligned}
\text{minimize} \quad & x_1 + x_2 + x_3 + x_4 + x_5 + x_6 \\
\text{subject to} \quad & x_1 \hspace{4.5cm} + x_6 \geq 65, \\
& x_1 + x_2 \hspace{3.5cm} \geq 75, \\
& \quad\;\; x_2 + x_3 \hspace{2.6cm} \geq 65, \\
& \qquad\;\; x_3 + x_4 \hspace{1.7cm} \geq 55, \\
& \qquad\qquad\; x_4 + x_5 \hspace{0.8cm} \geq 25, \\
& \qquad\qquad\qquad\; x_5 + x_6 \geq 35, \\
& x_1, \quad x_2, \quad x_3, \quad x_4, \quad x_5, \quad x_6 \geq 0.
\end{aligned}
$$

4.3 Convex Sets

In order to know the concept of a convex set, we firstly understand about the line segment. Let $P(x_1, y_1)$ and $Q(x_2, y_2)$ be two points in $\mathbb{R}^2$ (i.e., two-dimensional space) as given in Figure 4.1. Equation of any line through these points is

$$z = \lambda x + (1 - \lambda)y, \quad \lambda \in [0, 1]$$

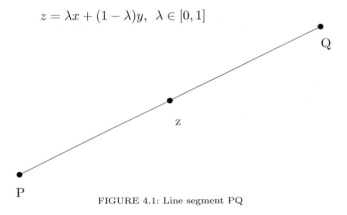

FIGURE 4.1: Line segment PQ

$$\frac{y - y_2}{y_1 - y_2} = \frac{x - x_2}{x_1 - x_2}.$$

Let

$$\frac{x - x_2}{x_1 - x_2} = \lambda.$$

That is,

$$x = \lambda x_1 + (1 - \lambda)x_2. \tag{4.1}$$

Similarly,

$$y = \lambda y_1 + (1 - \lambda)y_2. \tag{4.2}$$

Consider the three vectors

$$v = (x, y), \quad v_1 = (x_1, y_1), \quad v_2 = (x_2, y_2). \tag{4.3}$$

With the help of (4.3), (4.1) and (4.2) can be combined to get

$$v = \lambda v_1 + (1 - \lambda)v_2. \tag{4.4}$$

When $\lambda = 1$, we get

$$v = v_1 \quad i.e., \mathrm{P}.$$

When $\lambda = 0$, we get

$$v = v_2 \quad i.e., \mathrm{Q}.$$

Therefore, for the line segment PQ, we must have $0 \le \lambda \le 1$. This is also true for $\mathbb{R}^n$. The elements of this space are the n component vectors $v = \begin{bmatrix} v_1 & v_2 & \cdots & v_n \end{bmatrix}^T$

Definition 4.1 (Convex Set). A set Ω in n-dimensional $\mathbb{R}^n$ is said to be a convex set if any two points x and y in Ω, the line segment joining the two points is also in Ω.

In other words, if $x \in \Omega$ and $y \in \Omega$ and also if $z \in \Omega$ for all values of $\lambda \in [0, 1]$ and $z = \lambda x + (1 - \lambda)y$, then Ω is called a "convex set". For example, a triangle and its interior form a convex set. See Figure 4.2.

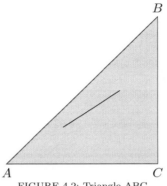

FIGURE 4.2: Triangle ABC

We can observe that Figure 4.3 is a convex set, but Figure 4.4 is not a convex set.

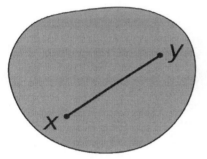

FIGURE 4.3: A convex set

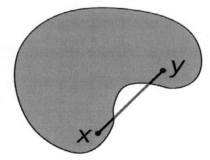

FIGURE 4.4: A nonconvex set

Example 4.5. Show that $\{(x, y): |x| \leq 5, |y| \leq 10\}$ is a convex set.

We Consider $X_1 = \{(x, y): |x| \leq 5, |y| \leq 10\}$. Let (x_1, y_1), $(x_2, y_2) \in X_1$. Then, $|x_1| \leq 5$, $|x_2| \leq 5$ and $|y_1| \leq 10$, $|y_2| \leq 10$. Suppose that $\lambda > 0$; therefore, we consider

$$\lambda(x_1, y_1) + (1 - \lambda)(x_2, y_2) = (\lambda x_1, \lambda y_1) + ((1 - \lambda)x_2, (1 - \lambda)y_2).$$

Next,

$$
\begin{aligned}
|\lambda x_1 + (1 - \lambda)x_2| &\leq |\lambda x_1| + |(1 - \lambda)x_2| \\
&= \lambda|x_1| + (1 - \lambda)|x_2| \\
&\leq \lambda 5 + (1 - \lambda)5 \\
&= 5\lambda + 5 - 5\lambda = 5.
\end{aligned}
$$

Therefore,

$$|\lambda x_1 + (1 - \lambda)x_2| \leq 5.$$

Similarly,

$$\begin{aligned}
|\lambda y_1 + (1 - \lambda)y_2| &\leq |\lambda y_1| + |(1 - \lambda)y_2| \\
&= \lambda|y_1| + (1 - \lambda)|y_2| \\
&\leq \lambda 10 + (1 - \lambda)10 \\
&= 10\lambda + 10 - 10\lambda = 10.
\end{aligned}$$

Therefore,

$$|\lambda y_1 + (1 - \lambda)y_2| \leq 10.$$

Thus, $\lambda(x_1, y_1) + (1 - \lambda)(x_2, y_2) \in X_1$ for $\lambda > 0$. Therefore, X_1 is a convex set.

4.4 Graphical Solution of Linear Programming Problem

We see graphically how linear programming optimizes a linear objective function in which the variables must satisfy a set of simultaneous linear equations. From the graphical view of points, we take following examples of linear programming problems of two variables and their analysis can be seen on a two-dimensional graph.

Example 4.6. Solve the following linear programming problem graphically.

$$\begin{aligned}
\text{maximize} \quad & 5x_1 + 7x_2 \\
\text{subject to} \quad & 3x_1 + 8x_2 \leq 12, \\
& x_1 + x_2 \leq 2, \\
& 2x_1 \qquad \leq 3, \\
& x_1, \quad x_2 \geq 0.
\end{aligned}$$

We consider the constraints as equalities

$$\begin{aligned}
3x_1 + 8x_2 &= 12, \\
x_1 + x_2 &= 2, \\
2x_1 &= 3.
\end{aligned}$$

See Figure 4.5.

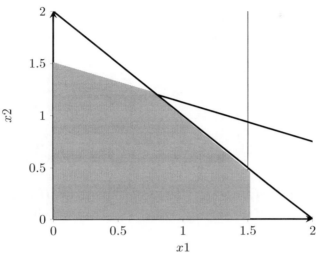

FIGURE 4.5: Graphical solution of Example 4.6

The maximum value of $5x_1 + 7x_2$ will be attainable at any one of the five vertices (extreme points) of the feasible region.

$$\text{At}(0,0), \quad 5 \times 0 + 7 \times 0 \qquad\qquad\qquad = 0.$$

$$\text{At}(\tfrac{3}{2},0), \quad 5 \times \tfrac{3}{2} + 7 \times 0 = \tfrac{15}{2} \qquad\qquad = 7.5.$$

$$\text{At}(\tfrac{3}{2},\tfrac{1}{2}), \quad 5 \times \tfrac{3}{2} + 7 \times \tfrac{1}{2} = \tfrac{15}{2} + \tfrac{7}{2} = \tfrac{22}{2} \quad = 11.$$

$$\text{At}(\tfrac{4}{5},\tfrac{6}{5}), \quad 5 \times \tfrac{4}{5} + 7 \times \tfrac{6}{5} = \tfrac{20}{5} + \tfrac{42}{5} = \tfrac{62}{5} \quad = 12.4.$$

$$\text{At}(0,\tfrac{3}{2}), \quad 5 \times 0 + 7 \times \tfrac{3}{2} = \tfrac{21}{2} \qquad\qquad = 10.5.$$

Thus, the objective function $5x_1 + 7x_2$ is maximum at $(\tfrac{4}{5}, \tfrac{6}{5})$.

In the Command Window,

```
>> mupad
k:=[{3*x1+8*x2<=12,x1+x2<=2,2*x1<=3},5*x1+7*x2,
    NonNegative]:
g:=linopt :: plot_data(k,[x1,x2]):
plot(g):
```

See Figure 4.6 as output in MATLAB.

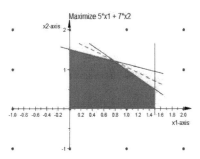

FIGURE 4.6: Graphical solution of Example 4.6 in MATLAB

Example 4.7. (*Production Planning in the Automobile Industry*) An automobile assembly plant assembles two types of vehicles: a four-door saloon and a people carrier. Both vehicle types must pass through a painting plant and an assembly plant. If the painting plant only paints four-door saloons, it can paint some 2,000 vehicles each day, whereas if it paints only people carriers, it can paint some 1,500 vehicles each day. Moreover, if the assembly plant only assembles either four-door saloons or people carriers, it can assemble some 2,200 vehicles every day. Each people carrier implies an average profit of \$3,000, whereas a four-door saloon implies an average profit of \$2,100.

(a) Use linear programming and indicate the daily production plan that would maximize the vehicle assembly plants daily profit.
Use linear programming and indicate the daily production plan that would maximize the vehicle assembly plant's daily profit.

Decision variables:
x_1=Hundreds of four-door saloons produced daily,
x_2=Hundreds of people carriers produced daily.

Objective function:
To maximize $Z = 21x_1 + 30x_2$.

Constraints:

$R1$: The fraction of the day during which the painting plant occupied is equal to or less than 1.

$R1$: The fraction of the day during which the painting plant works on four-door saloons: $\frac{1}{2.000}$.

$R1$: The fraction of the day during which the painting plant works on people carriers: $\frac{1}{1.500}$.

$R1$: $\frac{1}{20}x_1 + \frac{1}{15}x_2 \leq 1$.

$R2$: The fraction of the day during which the assembly plant occupied is equal to or less than 1.

$R2$: The fraction of the day during which the assembly plant works on four-door saloons or people carriers: $\frac{1}{2.200}$.

$R2$: $\frac{1}{22}x_1 + \frac{1}{22}x_2 \leq 1$.

$R3$: The non-negativity constraint.

$R3$: $x_1, x_2 \geq 0$.

As this model has two decision variables, the problem can be solved graphically which is shown in Figure 4.7.

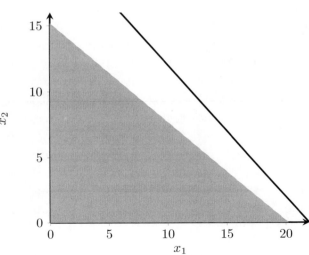

FIGURE 4.7: Graphical solution of Example 4.7

In the Command Window
>> mupad

k:=[{1/20*x1+1/15*x2<=1,1/22*x1+1/22*x2<=1},
            21*x1+30*x2, NonNegative]:
g:=linopt :: plot_data(k,[x1,x2]):
plot(g):

Output: See Figure 4.8 as output in MATLAB.

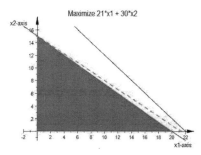

FIGURE 4.8: Graphical solution of Example 4.7 in MATLAB

(b) What surpluses would be produced in the painting plant and the assembly plant?
The painting plant is saturated, that is, it has no surplus, yet the assembly plant has a surplus of 0.3.

Example 4.8. (*Investment of Funds*) A small investor has $12,000 to invest and three different funds to choose from. Guaranteed investment funds offer an expected rate of return of 7%, mixed funds (part is guaranteed capital) have an expected rate of return of 8%, while an investment on the Stock Exchange involves an expected rate of return of 12%, but without guaranteed investment capital. In order to minimize the risk, the investor has decided to not invest more than $2,000 on the Stock Exchange. Moreover, for tax reasons, the investor needs to invest at least three times more in guaranteed investment funds than in mixed funds. Let us assume that at the end of the year the returns are those expected; what are the optimum investment amounts?

(a) Consider this problem as if it were a linear programming model with two decision variables.

Decision variables:

x_1: amount (in thousands of $\$$) invested in guaranteed funds;

x_2: amount (in thousands of $\$$) invested in mixed funds;

Objective function:

$$\text{maximize} \quad z = 0.07x_1 + 0.08x_2 + 0.12(12 - x_1 - x_2)$$
$$= 1.44 - 0.05x_1 - 0.04x_2.$$

Constraints:

The non-negativity constraint of the amount invested,

$$12 - x_1 - x_2 \geq 0.$$

Upper limit of the amount invested,

$$(12 - x_1 - x_2) \leq 2.$$

Constraint for tax reasons,

$$x_2 \leq \frac{1}{3}x_1,$$

where

$$x_1, x_2 \geq 0.$$

(b) Solve the problem with the graphic method and indicate the optimum solution.

The feasible region is given in Figure 4.9.

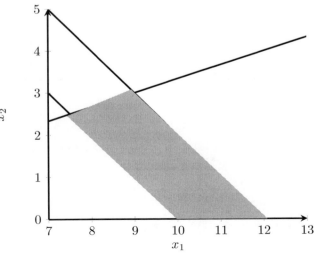

FIGURE 4.9: Graphical solution of Example 4.8

Corner points are (9, 3), (12, 0), (10, 0) and (7.5, 2.5); therefore, we get an optimum investment of $965.

In the Command Window,

`>> mupad`

k:=[{12−x1−x2>=0,12−x1−x2<=2,x2<=0.33∗x1},1.44−
                0.05∗x1−0.04∗x2, NonNegative]:
g:=linopt :: plot_data(k,[x1,x2]):
plot(g):

Output: See Figure 4.10 as output in MATLAB.

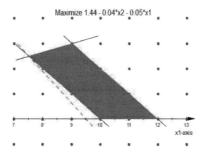

FIGURE 4.10: Graphical solution of Example 4.8 in MATLAB

Example 4.9. Solve graphically.

$$\begin{aligned}
\text{minimize} \quad & 4x_1 + 2x_2 \\
\text{subject to} \quad & x_1 + 2x_2 \geq 2, \\
& 3x_1 + x_2 \geq 3, \\
& 4x_1 + 3x_2 \geq 6, \\
& x_1, \quad x_2 \geq 0.
\end{aligned}$$

`>> mupad`

k:=[{x1+2∗x2>=2,3∗x1+x2>=3,4∗x1+3∗x2>=6},
      4∗x1+2∗x2, NonNegative]:
g:=linopt :: plot_data(k,[x1,x2]):
plot(g)

Output: See Figure 4.11 as output in MATLAB.

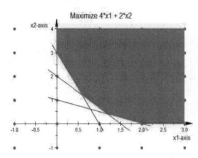

FIGURE 4.11: Graphical solution of Example 4.9 in MATLAB

An optimum value of the objective function is $\frac{24}{5}$ at $\left(\frac{3}{5}, \frac{6}{5}\right)$.

Example 4.10. For the linear programming problem,

$$
\begin{aligned}
\text{minimize} \quad & x_1 - x_2 \\
\text{subject to} \quad & 2x_1 + 3x_2 \le 6, \\
& x_1 \qquad \le 3, \\
& \qquad x_2 \le 3, \\
& x_1, \qquad x_2 \ge 0.
\end{aligned}
$$

The number of corner points are $x_1 = x_2 = 0$; $x_1 = 3, x_2 = 0$ and $x_1 = 0, x_2 = 2$. They are shown in Figure 4.12.

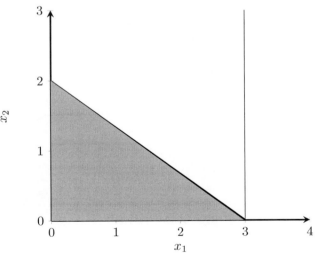

FIGURE 4.12: Graphical solution of Example 4.10

We can draw the graph in MATLAB.

In the Command Window,

```
>> mupad
k:=[{2*x1+3*x2<=6,x1<=3,x2<=3},x1−x2,
                        NonNegative]:
g:=linopt::plot_data(k,[x1,x2]):
plot(g):
```

See Figure 4.13 as output in MATLAB.

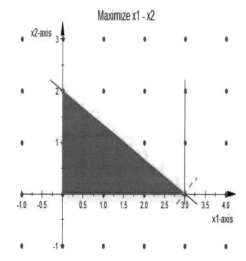

FIGURE 4.13: Graphical solution of Example 4.10 in MATLAB

Example 4.11. Solve the following linear programming problem by graphical method.

$$\begin{aligned}
\text{maximize} \quad & x_1 + x_2 \\
\text{subject to} \quad & 2x_1 + x_2 \geq 8, \\
& 2x_1 + 5x_2 \geq 10, \\
& x_1, x_2 \geq 0.
\end{aligned}$$

The optimal value of this problem is_____.
See Figure 4.14.

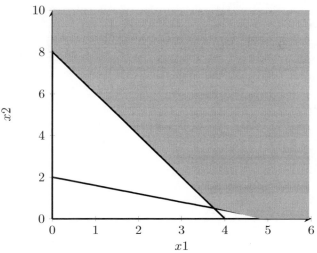

FIGURE 4.14: Graphical solution of Example 4.11

Corner points are $(0, 8)$, $(\frac{15}{4}, \frac{1}{2})$ and $(5, 0)$. The optimal solution is found at $(0, 8)$ and the maximum value of an objective function is 8.

In the Command Window,

```
>> mupad
k:=[{2*x1+x2>=8,2*x1+5*x2>=10},x1+x2,
                          NonNegative]:
g:=linopt::plot_data(k,[x1,x2]):
plot(g):
```

See Figure 4.15 as output in MATLAB.

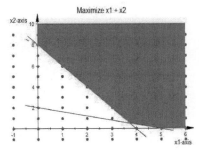

FIGURE 4.15: Graphical solution of Example 4.11 in MATLAB

Example 4.12. Solve the following linear program graphically using MATLAB.

$$\text{maximize} \quad 2x_1 + 5x_2$$
$$\text{subject to} \quad x_1 \qquad \leq 4,$$
$$x_2 \leq 6,$$
$$x_1 + \ x_2 \leq 8,$$
$$x_1, \quad x_2, \geq 0.$$

In the Command Window,

```
>> mupad
k:=[{x1<=4,x2<=6,x1+x2<=8},2*x1+5*x2,
                        NonNegative]:
g:=linopt::plot_data(k,[x1,x2]):
plot(g):
```

Output: See Figure 4.16 as output in MATLAB.

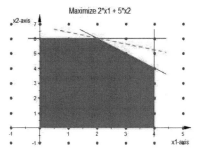

FIGURE 4.16: Graphical solution of Example 4.12 in MATLAB

Corner points are (4,0), (4,4), (2,6) and (0,6). It clearly shows that the maximum value of an objective function at $(2,6)$ is $2x_1 + 5x_2 = 2 \times 2 + 5 \times 6 = 34$.

Example 4.13. Consider the linear programming problem

$$\text{maximize} \quad x_1 + \ x_2$$
$$\text{subject to} \quad x_1 - 2x_2 \leq 10,$$
$$-2x_1 + \ x_2 \leq 10,$$
$$x_1, \quad x_2, \geq \ 0.$$

Then, which of the following options is TRUE?

(a) The linear programming problem admits an optimal solution.

(b) The linear programming problem is unbounded.

(c) The linear programming problem admits no feasible solution.

(d) The linear programming problem admits a unique feasible solution.

In the Command Window,

```
k:=[{x1−2*x2<=10,x2−2*x1<=10},x1+x2 ,
                                NonNegative]:
g:=linopt :: plot_data (k ,[x1 ,x2 ]):
plot(g):
```

See Figure 4.17 as output in MATLAB.

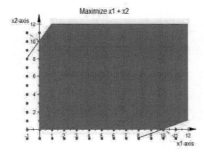

FIGURE 4.17: Graphical solution of Example 4.13 in MATLAB

From the above graph in Figure 4.17, we observe that the linear programming problem has unbounded solution. Option (b) is ture.

Example 4.14. Suppose that the variables $x_1 \geq 0$ and $x_2 \geq 0$ satisfy the constraints $x_1 + x_2 \geq 3$ and $x_1 + 2x_2 \geq 4$. Which of the following is true?

(a) The maximum value of $5x_1 + 7x_2$ is 231 and it does not have any finite minimum.

(b) The minimum value of $5x_1 + 7x_2$ is 17 and it does not have any finite maximum.

(c) The maximum value of $5x_1 + 7x_2$ is 231 and its minimum value is 17.

(d) $5x_1 + 7x_2$ neither has a finite maximum nor a finite minimum.

See Figure 4.18.

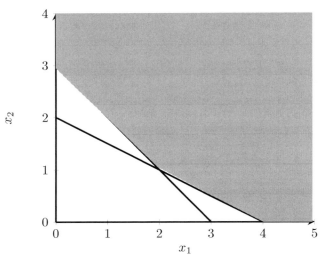

FIGURE 4.18: Graphical solution of Example 4.14

$$\text{At}(0,3), \quad 5x_1 + 7x_2 = 5 \times 0 + 7 \times 3 = 21.$$
$$\text{At}(2,1), \quad 5x_1 + 7x_2 = 5 \times 2 + 7 \times 1 = 17.$$
$$\text{At}(4,0), \quad 5x_1 + 7x_2 = 5 \times 4 + 7 \times 0 = 20.$$

In the Command Window,

```
>> mupad
```

$$k := [\{x1+x2>=3, x1+2*x2>=4\}, 5*x1+7*x2, \text{NonNegative}]:$$
$$g := \text{linopt}::\text{plot\_data}(k, [x1, x2]):$$
$$\text{plot}(g):$$

See Figure 4.19 as output in MATLAB.

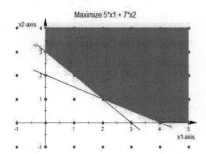

FIGURE 4.19: Graphical solution of Example 4.14 in MATLAB

The minimum value of $5x_1 + 7x_2$ is 17 and it does not have any finite maximum. Therefore, option (b) is true.

4.5   Exercises

Exercise 4.1. A production company produces two types of models: M_1 and M_2. Each M_1 model requires 4 hours of grinding and 2 hours of polishing; whereas, each M_2 model requires 2 hours of grinding and 5 hours of polishing. The company has 2 grinders and 3 polishers. Each grinder works for 40 hours per week and each polisher works for 60 hours per week. Profit on an M_1 model is \$3 and on an M_2 model is \$4. Whatever produced in a week is sold in the market. How should the production company allocate his production capacity to the two types of models so that he may make maximum profit in a week.

Exercise 4.2. On the bank of Ganga river, Varanasi, there are three neighbouring cities that are discharging two kinds of pollutants, A and B, into the river. Now the Uttar Pradesh state government has set up a treatment plant that treats pollutants from City 1 for \$15 per ton which reduces pollutants A and B by the amount of 0.10 and 0.45 tons per ton of waste, respectively. It costs \$10 per ton to process a ton of City 2 waste and consequentially reducing pollutants A and B by 0.20 and 0.25 tons per ton of waste, respectively. Similarly, City 3 waste is treated for \$20 reducing A by 0.40 and B by 0.30 tons per ton of waste. The state wishes to reduce the amount of pollutant A by at least 30 and B by 40 tons. Formulate the linear programming problem that will minimize the cost of reducing pollutants by the desired amount.

Exercise 4.3. Which ones of the following subsets of $\mathbb{R}^2$ are convex and which are not?

(a) $\{(x, y): x^2 + y^2 = 1\}$

(b) $\{(x, y): y \geq x^2\}$

(c) $\{(x, y): y \leq x^2\}$

(d) $\{(x, y): x^2 + y^2 \leq 4\}$

(e) $\{(x, y): x^2 + y^2 \geq 3\}$

Exercise 4.4. Solve the following linear programming problems by graphical method.

(a)

$$\begin{aligned}
\text{maximize} \quad & 3x_1 + 2x_2 \\
\text{subject to} \quad & 2x_1 - x_2 \geq 2, \\
& x_1 + 2x_2 \leq 8, \\
& x_1, \quad x_2 \geq 0.
\end{aligned}$$

(b)

$$\begin{aligned}
\text{maximize} \quad & 2x_1 + 3x_2 \\
\text{subject to} \quad & x_1 + x_2 \leq 1, \\
& 3x_1 + x_2 \leq 4, \\
& x_1, \quad x_2 \geq 0.
\end{aligned}$$

(c)

$$\begin{aligned}
\text{maximize} \quad & 3x_1 + 2x_2 \\
\text{subject to} \quad & 2x_1 + x_2 \leq 1, \\
& x_1 \quad\quad \leq 2, \\
& x_1 + x_2 \geq 3, \\
& x_1, \quad x_2 \geq 0.
\end{aligned}$$

(d)

$$\begin{aligned}
\text{maximize} \quad & 3x_1 + 4x_2 \\
\text{subject to} \quad & x_1 - x_2 \leq -1, \\
& -x_1 + x_2 \leq 0, \\
& x_1, \quad x_2 \geq 0.
\end{aligned}$$

(e)

$$\begin{aligned}
\text{maximize} \quad & 8000x_1 + 7000x_2 \\
\text{subject to} \quad & 3x_1 + x_2 \leq 66, \\
& x_1 + x_2 \leq 45, \\
& x_1 \quad\quad \leq 20, \\
& x_2 \leq 40, \\
& x_1, \quad x_2 \geq 0.
\end{aligned}$$

Exercise 4.5. Solve the following linear programming problems graphically.

(a)

$$\begin{aligned}
\text{maximize} \quad & 5x_1 + 4x_2 \\
\text{subject to} \quad & x_1 - 2x_2 \leq 1, \\
& x_1 + 2x_2 \leq 3, \\
& x_1, \quad x_2 \geq 0.
\end{aligned}$$

(b)

$$\text{maximize} \quad 3x_1 + 2x_2$$
$$\text{subject to} \quad 3x_1 - 2x_2 \geq -20,$$
$$-2x_1 + 3x_2 \leq 9,$$
$$x_1, \quad x_2 \geq 0.$$

Exercise 4.6. Solve the following linear programming problem graphically.

$$\text{maximize} \quad 100x_1 + 100x_2$$
$$\text{subject to} \quad 10x_1 + 5x_2 \leq 80,$$
$$6x_1 + 6x_2 \leq 66,$$
$$4x_1 + 8x_2 \geq 24,$$
$$5x_1 + 6x_2 \leq 90,$$
$$x_1, \quad x_2 \geq 0.$$

Chapter 5

The Simplex Method

5.1 Standard Form of Linear Programming Problem

We refer to a linear programming problem in standard form as follows:

$$\begin{array}{ll} \text{minimize} & c^T x \\ \text{subject to} & Ax = b, \\ \text{where} & c \in \mathbb{R}^n, x \in \mathbb{R}^n, b \in \mathbb{R}^m, A \in \mathbb{R}^{m \times n}, x \geq 0. \end{array}$$

The above problem can also be written as follows:

minimize $\quad c_1 x_1 + c_2 x_2 + \cdots + c_n x_n$

subject to

$$\begin{bmatrix} a_{11} & a_{12} & \ldots & a_{1n} \\ a_{21} & a_{22} & \ldots & a_{2n} \\ \vdots & \vdots & \ldots & \vdots \\ a_{m1} & a_{m2} & \ldots & a_{mn} \end{bmatrix} \begin{bmatrix} x_1 \\ x_2 \\ \vdots \\ x_n \end{bmatrix} = \begin{bmatrix} b_1 \\ b_2 \\ \vdots \\ b_m \end{bmatrix},$$

$$\begin{bmatrix} x_1 \\ x_2 \\ \vdots \\ x_n \end{bmatrix} \geq \begin{bmatrix} 0 \\ 0 \\ \vdots \\ 0 \end{bmatrix}.$$

Equivalently.

$$\begin{array}{ll} \text{minimize} & c_1 x_1 + c_2 x_2 + \cdots + c_n x_n \\ \text{subject to} & a_{11} x_1 + a_{12} x_2 + \cdots + a_{1n} x_n = b_1, \\ & a_{21} x_1 + a_{22} x_2 + \cdots + a_{2n} x_n = b_2, \\ & \quad\quad\quad\quad\quad \vdots \\ & a_{m1} x_1 + a_{m2} x_2 + \cdots + a_{mn} x_n = b_m, \end{array}$$

where c_i, b_i and a_{ij} are fixed real constants and x_i is the real number to be determined.

5.2   Basic Solutions

Consider a linear programming problem in standard form:

minimize $c^T x$,
subject to $Ax = b$,
where $x \geq 0, c \in \mathbb{R}^n, x \in \mathbb{R}^n, b \in \mathbb{R}^m, A \in \mathbb{R}^{m \times n}, m < n$.

We consider the system of equations as $Ax = b$ where rank $A = m$. Let B be a square matrix of order $m \times m$ whose columns are linearly independent columns of the matrix A. We can write matrix A in the form $\begin{bmatrix} B \colon D \end{bmatrix}$ where $D = m \times (n - m)$ whose columns are remaining columns of A. The matrix B is nonsingular. We can solve

$$Bx_B = b,$$

that is

$$x_B = B^{-1}b.$$

Thus, $x = \begin{bmatrix} x_B^T, 0^T \end{bmatrix}^T$ is a solution of $Ax = b$.

Definition 5.1 (Basic Solution). We say $x = \begin{bmatrix} x_B^T, 0^T \end{bmatrix}^T$ is a basic solution of $Ax = b$ with respect to the basis matrix B. The components of x_B are called "basic variables". If some of the basic variables of a basic solution are zero, then the basic solution is called a "degenerate basic solution".

Definition 5.2 (Feasible Solution). A vector $x \in \mathbb{R}^n$ satisfying $Ax = b$, where $x \geq 0$ is called a "feasible solution". A feasible solution that is also basic is called a "basic feasible solution".

Definition 5.3 (Degenerate Basic Feasible Solution). If the basic feasible solution is a degenerate basic solution, then it is called a "degenerate basic feasible solution". We must note that $x_B \geq 0$ for any basic feasible solution.

Thus, we solve the following linear equations to find basic solutions and also check the feasible solution.

Example 5.1. Find all basic solutions of the following equations:

$$\begin{aligned} x_1 + \ x_2 + x_3 &= 1, \\ 2x_1 + 3x_2 \ \ \ &= 1. \end{aligned}$$

Augmented matrix is

$$\begin{bmatrix} 1 & 1 & 1 & | & 1 \\ 2 & 3 & 0 & | & 1 \end{bmatrix} \sim \begin{bmatrix} 1 & 1 & 0 & | & 1 \\ 0 & 1 & -2 & | & -1 \end{bmatrix} \sim \begin{bmatrix} 1 & 0 & 2 & | & 2 \\ 0 & 1 & -2 & | & -1 \end{bmatrix}$$

$$R_2 \to R_2 - 2R_1 \qquad R_1 \to R_1 - R_2$$

System of linear equations can be given as

$$\begin{aligned} x_1 \quad + 2x_3 &= \ \ 2, \\ x_2 \quad - 2x_3 &= -1. \end{aligned}$$

Solving the above two equations,

$$x_1 = 2 - 2s,$$
$$x_2 = -1 + 2s,$$
$$x_3 = s, \ s \in \mathbb{R}.$$

Thus, the system of linear equations has infinitely many solutions, but it can have at most $\binom{3}{2}$, i.e., 3 basic solutions.

1. Since $B = \begin{bmatrix} a_1 & a_2 \end{bmatrix}$ is a basis matrix, then we have $x_B = \begin{bmatrix} x_1 \\ x_2 \end{bmatrix}$.

    Therefore, $Bx_B = b$ and its augmented matrix is

    $$\begin{bmatrix} 1 & 1 & | & 1 \\ 2 & 3 & | & 1 \end{bmatrix} \sim \begin{bmatrix} 1 & 1 & | & 1 \\ 0 & 1 & | & -1 \end{bmatrix}.$$

    $$R_2 \to R_2 - 2R_1$$

    Therefore,

    $$\begin{aligned} x_1 + x_2 &= 1, \\ x_2 &= -1. \end{aligned}$$

Therefore, the basic vector, i.e., $x_1 = 2$, $x_2 = -1$, $x_B = \begin{bmatrix} 2 \\ -1 \end{bmatrix}$

and $x = \begin{bmatrix} 2 \\ -1 \\ 0 \end{bmatrix}$. It is not a basic feasible solution.

2. Since basis matrix is $B = \begin{bmatrix} a_1 & a_3 \end{bmatrix}$, therefore, $x_B = \begin{bmatrix} x_1 \\ x_3 \end{bmatrix}$. We write

$$\left[\begin{array}{cc|c} 1 & 1 & 1 \\ 2 & 0 & 1 \end{array} \right].$$

$$2x_1 = 1,$$
$$x_1 + x_3 = 1.$$

The basic vector, i.e., $x_1 = \frac{1}{2}$, $x_3 = \frac{1}{2}$, $x_B = \begin{bmatrix} 1/2 \\ 1/2 \end{bmatrix}$ and $x = \begin{bmatrix} 1/2 \\ 0 \\ 1/2 \end{bmatrix}$. It is a basic feasible solution.

3. Since basis matrix is $B = \begin{bmatrix} a_2 & a_3 \end{bmatrix}$, therefore $x_B = \begin{bmatrix} x_2 \\ x_3 \end{bmatrix}$. We write

$$\left[\begin{array}{cc|c} 1 & 1 & 1 \\ 3 & 0 & 1 \end{array} \right].$$

$$3x_2 = 1,$$
$$x_2 + x_3 = 1.$$

Therefore, basic vector, i.e., $x_2 = \frac{1}{3}$, $x_3 = \frac{2}{3}$, $x_B = \begin{bmatrix} 1/3 \\ 2/3 \end{bmatrix}$ and $x = \begin{bmatrix} 0 \\ 1/3 \\ 2/3 \end{bmatrix}$. It is a basic feasible solution.

Example 5.2. Consider the equation $Ax = b$, where $A =$

$$\begin{bmatrix} 2 & 6 & 2 & 1 \\ 6 & 4 & 4 & 6 \end{bmatrix}, \ b = \begin{bmatrix} 3 \\ 2 \end{bmatrix} \text{ and } x = \begin{bmatrix} x_1 \\ x_2 \\ x_3 \\ x_4 \end{bmatrix}. \text{ Find basic solutions of}$$

$Ax = b.$

The augmented matrix

$$\begin{aligned}
[\ A\mid b\ ] &= \begin{bmatrix} 2 & 6 & 2 & 1 & 3 \\ 6 & 4 & 4 & 6 & 2 \end{bmatrix} \quad R_1 \to \frac{1}{2}R_1 \\[2mm]
&\sim \begin{bmatrix} 1 & 3 & 1 & 1/2 & 3/2 \\ 6 & 4 & 4 & 6 & 2 \end{bmatrix} \quad R_2 \to R_2 - 6R_1 \\[2mm]
&\sim \begin{bmatrix} 1 & 3 & 1 & 1/2 & 3/2 \\ 0 & -14 & -2 & 3 & -7 \end{bmatrix} \quad R_2 \to -\frac{1}{14}R_2 \\[2mm]
&\sim \begin{bmatrix} 1 & 3 & 1 & 1/2 & 3/2 \\ 0 & 1 & 1/7 & -3/14 & 1/2 \end{bmatrix} \quad R_1 \to R_1 - 3R_2 \\[2mm]
&\sim \begin{bmatrix} 1 & 0 & 4/7 & 8/7 & 0 \\ 0 & 1 & 1/7 & -3/14 & 1/2 \end{bmatrix}.
\end{aligned}$$

Corresponding system of linear equations is given by

$$x_1 + \frac{4}{7}x_3 + \frac{8}{7}x_4 = 0,$$

$$x_2 + \frac{1}{7}x_3 - \frac{3}{14}x_4 = \frac{1}{2}.$$

Solving for unknown variables x_1 and x_2, we get

$$x_1 = -\frac{4}{7}x_3 - \frac{8}{7}x_4,$$

$$x_2 = \frac{1}{2} - \frac{1}{7}x_3 + \frac{3}{14}x_4.$$

Taking $x_3 = s \in \mathbb{R}$ and $x_4 = t \in \mathbb{R}$, we get

$$x_1 = -\frac{4}{7}s - \frac{8}{7}t,$$

$$x_2 = \frac{1}{2} - \frac{1}{7}s + \frac{3}{14}t,$$

$$x_3 = s,$$

$$x_4 = t.$$

In vector notation, we may write the system of equations above as

$$\begin{bmatrix} x_1 \\ x_2 \\ x_3 \\ x_4 \end{bmatrix} = \begin{bmatrix} 0 \\ 1/2 \\ 0 \\ 0 \end{bmatrix} + s \begin{bmatrix} -4/7 \\ -1/7 \\ 1 \\ 0 \end{bmatrix} + t \begin{bmatrix} -8/7 \\ 3/14 \\ 0 \\ 1 \end{bmatrix}.$$

We have infinitely many solutions for $s, t \in \mathbb{R}$. Our question is how many basic feasible solutions are there?

We have unknown variables $m = 4$ and number of equations $n = 2$.

$$\binom{m}{n} = \binom{4}{2} = \frac{4!}{2!2!} = 6.$$

Therefore, we have at most six basic feasible solutions. We try to check each of the basic solutions for feasibility.

1. Since basis matrix $B = \begin{bmatrix} 2 & 6 \\ 6 & 4 \end{bmatrix}$, then $x_B = \begin{bmatrix} x_1 \\ x_2 \end{bmatrix}$. We solve $Bx_B = b$.

$$\begin{bmatrix} 2 & 6 & 3 \\ 6 & 4 & 2 \end{bmatrix} \sim \begin{bmatrix} 2 & 6 & 3 \\ 0 & -14 & -7 \end{bmatrix}.$$

$$R_2 \to R_2 - 3R_1$$

The corresponding system of linear equations is

$$2x_1 + 6x_2 = 3,$$
$$-14x_2 = -7.$$

After solving the above two equations, we get basic vector,

i.e., $x_1 = 0$, $x_2 = \frac{1}{2}$, $x_B = \begin{bmatrix} 0 \\ 1/2 \end{bmatrix}$, and $x = \begin{bmatrix} 0 \\ 1/2 \\ 0 \\ 0 \end{bmatrix}$ is a solution of $Ax = b$. We observe that x is a basic solution of $Ax = b$. Since $x \geq 0$, it is also feasible.

2. Since basis $B = \begin{bmatrix} 2 & 2 \\ 6 & 4 \end{bmatrix}$, therefore $x_B = \begin{bmatrix} x_1 \\ x_3 \end{bmatrix}$. We solve $Bx_B = b$.

$$\begin{bmatrix} 2 & 2 & | & 3 \\ 6 & 4 & | & 2 \end{bmatrix} \sim \begin{bmatrix} 1 & 1 & | & 3/2 \\ 6 & 4 & | & 2 \end{bmatrix} \sim \begin{bmatrix} 1 & 1 & | & 3/2 \\ 0 & -2 & | & -7 \end{bmatrix}$$
$$R_1 \to \tfrac{1}{2}R_1 \qquad R_2 \to R_2 - 6R_1 \qquad R_2 \to -\tfrac{1}{2}R_2$$

$$\sim \begin{bmatrix} 1 & 1 & | & 3/2 \\ 0 & 1 & | & 7/2 \end{bmatrix} \sim \begin{bmatrix} 1 & 0 & | & -2 \\ 0 & 1 & | & 7/2 \end{bmatrix}.$$
$$R_1 \to R_1 - R_2$$

We get $x_1 = -2$ and $x_3 = \tfrac{7}{2}$. Thus, $x_B = \begin{bmatrix} -2 \\ 7/2 \end{bmatrix}$ and $x = \begin{bmatrix} -2 \\ 0 \\ 7/2 \\ 0 \end{bmatrix}$ is a solution of $Ax = b$. It is basic but it is not feasible.

3. Since basis matrix is $B = \begin{bmatrix} 2 & 1 \\ 6 & 6 \end{bmatrix}$, therefore, $x_B = \begin{bmatrix} x_1 \\ x_4 \end{bmatrix}$. We solve $Bx_B = b$.

$$\begin{bmatrix} 2 & 1 & | & 3 \\ 6 & 6 & | & 2 \end{bmatrix} \sim \begin{bmatrix} 2 & 1 & | & 3 \\ -6 & 0 & | & -16 \end{bmatrix} \sim \begin{bmatrix} 2 & 1 & | & 3 \\ 1 & 0 & | & 8/3 \end{bmatrix}$$
$$R_2 \to R_2 - 6R_1 \qquad R_2 \to -\tfrac{1}{6}R_2 \qquad R_1 \to R_1 - 2R_2$$

$$\sim \begin{bmatrix} 0 & 1 & | & -7/3 \\ 1 & 0 & | & 8/3 \end{bmatrix} \sim \begin{bmatrix} 1 & 0 & | & 8/3 \\ 0 & 1 & | & -7/3 \end{bmatrix}.$$
$$R_1 \leftrightarrow R_2$$

We get $x_1 = \tfrac{8}{3}$ and $x_4 = -\tfrac{7}{3}$. Thus, $x_B = \begin{bmatrix} 8/3 \\ -7/3 \end{bmatrix}$ and

$$x = \begin{bmatrix} 8/3 \\ 0 \\ 0 \\ -7/3 \end{bmatrix}$$ is a solution of $Ax = b$. It is basic, but it is not feasible.

4. Since basis matrix $B = \begin{bmatrix} 6 & 2 \\ 4 & 4 \end{bmatrix}$, therefore $x_B = \begin{bmatrix} x_2 \\ x_3 \end{bmatrix}$. We solve

$Bx_B = b.$

$$\begin{bmatrix} 6 & 2 & | & 3 \\ 4 & 4 & | & 2 \end{bmatrix} \sim \begin{bmatrix} 6 & 2 & | & 3 \\ 1 & 1 & | & 1/2 \end{bmatrix} \sim \begin{bmatrix} 0 & -4 & | & 0 \\ 1 & 1 & | & 1/2 \end{bmatrix}$$

$\quad R_2 \to \frac{1}{4}R_2 \qquad R_1 \to R_1 - 6R_2 \qquad R_1 \to -\frac{1}{4}R_1$

$$\sim \begin{bmatrix} 0 & 1 & | & 0 \\ 1 & 1 & | & 1/2 \end{bmatrix}.$$

Thus, we get $x_3 = 0$, $x_2 = \frac{1}{2}$ and $x = \begin{bmatrix} 0 \\ 1/2 \\ 0 \\ 0 \end{bmatrix}$. It gives a basic

feasible solution.

5. Since $B = \begin{bmatrix} 6 & 1 \\ 4 & 6 \end{bmatrix}$ and $x_B = \begin{bmatrix} x_2 \\ x_4 \end{bmatrix}$, then we solve $Bx_B = b$.

$$\begin{bmatrix} 6 & 1 & | & 3 \\ 4 & 6 & | & 2 \end{bmatrix} \sim \begin{bmatrix} 6 & 1 & | & 3 \\ -32 & 0 & | & -16 \end{bmatrix} \sim \begin{bmatrix} 6 & 1 & | & 3 \\ 1 & 0 & | & 1/2 \end{bmatrix}$$

$R_2 \to R_2 - 6R_1 \qquad R_2 \to -\frac{1}{32}R_2 \qquad R_1 \to R_1 - 6R_2$

$$\sim \begin{bmatrix} 0 & 1 & | & 0 \\ 1 & 0 & | & 1/2 \end{bmatrix} \sim \begin{bmatrix} 1 & 0 & | & 1/2 \\ 0 & 1 & | & 0 \end{bmatrix}.$$

$\quad R_1 \leftrightarrow R_2$

We get $x_2 = \frac{1}{2}$, $x_4 = 0$ and $x = \begin{bmatrix} 0 \\ 1/2 \\ 0 \\ 0 \end{bmatrix}$. It is basic feasible

solution.

6. Since basis matrix is $B = \begin{bmatrix} 2 & 1 \\ 4 & 6 \end{bmatrix}$; therefore, basic vector is

$x_B = \begin{bmatrix} x_3 \\ x_4 \end{bmatrix}$. Then, we solve $Bx_B = b$.

$$\begin{bmatrix} 2 & 1 & | & 3 \\ 4 & 6 & | & 2 \end{bmatrix} \sim \begin{bmatrix} 2 & 1 & | & 3 \\ -8 & 0 & | & -16 \end{bmatrix} \sim \begin{bmatrix} 2 & 1 & | & 3 \\ 1 & 0 & | & 2 \end{bmatrix}$$

$R_2 \to R_2 - 6R_1 \qquad R_2 \to -\frac{1}{8}R_2 \qquad R_1 \to R_1 - 2R_2$

$$\sim \begin{bmatrix} 0 & 1 & | & -1 \\ 1 & 0 & | & 2 \end{bmatrix} \sim \begin{bmatrix} 1 & 0 & | & 2 \\ 0 & 1 & | & -1 \end{bmatrix}.$$

$\quad R_1 \leftrightarrow R_2$

We get $x_3 = 2$, $x_4 = -1$ and $x = \begin{bmatrix} 0 \\ 0 \\ 2 \\ -1 \end{bmatrix}$. It is basic but not feasible.

5.3   Properties of Basic Solutions

In this section, we discuss the importance of basic feasible solutions in solving linear programming problems. We prove the fundamental theorem of a linear programming problem. Before this, we should have an idea about an optimal basic feasible solution.

Definition 5.4 (Optimal Basic Feasible Solution). Any vector x that yields the minimum value of the objective function $c^T x$ over the set of vectors satisfying $Ax = b$, $x \geq 0$ is called an "optimal feasible solution". An optimal feasible solution that is basic is called an "optimal basic feasible solution".

Theorem 5.1. *The set of feasible solutions of a standard form linear programming problem is a convex set.*

Proof. Consider a standard form of linear programming problem:

$$\begin{aligned} &\text{minimize} \quad c^T x, \\ &\text{subject to} \quad Ax = b, \\ &\text{where} \quad c \in \mathbb{R}^n, x \in \mathbb{R}^n, b \in \mathbb{R}^m, A \in \mathbb{R}^{m \times n}, x \geq 0. \end{aligned}$$

The set of feasible solution is $X := \{X \in \mathbb{R}^n \colon Ax = b, x \geq 0\}$. We have to prove that X is a convex set. That is, if $x, y \in X$ and $\lambda \in [0, 1]$, then we have to show that $\lambda x + (1 - \lambda)y \in X$. For that, we have to show that

$$A(\lambda x + (1 - \lambda)y) = b \text{ and } \lambda x + (1 - \lambda y) \geq 0.$$

Since

$$x, y \in X,$$

therefore

$$Ax = b,$$
$$Ay = b.$$

Then,

$$A(\lambda x + (1 - \lambda)y) = \lambda Ax + (1 - \lambda)Ay$$
$$= \lambda b + (1 - \lambda)b$$
$$= b.$$

That is,

$$A(\lambda x + (1 - \lambda)y) = b. \qquad (5.1)$$

Since

$$x, y \geq 0,$$

therefore

$$\lambda x + (1 - \lambda)y \geq \lambda \times 0 + (1 - \lambda) \times 0.$$

That is,

$$\lambda x + (1 - \lambda)y \geq 0. \qquad (5.2)$$

From (5.1) and (5.2),

$$\lambda x + (1 - \lambda)y \in X.$$

Thus, X is a convex set.   □

Theorem 5.2. [Fundamental Theorem of Linear Programming Problem] *Consider a linear programming problem in standard form:*

$$\begin{aligned} &minimize \quad c^T x, \\ &subject\ to \quad Ax = b, \\ &where \qquad c \in \mathbb{R}^n, x \in \mathbb{R}^n, b \in \mathbb{R}^m, A \in \mathbb{R}^{m \times n}, x \geq 0. \end{aligned}$$

1. *If there exists a feasible solution, then there exists a basic feasible solution;*

2. *If there exists an optimal feasible solution, then there exists an optimal basic feasible solution.*

Proof. 1. Suppose that $x = \begin{bmatrix} x_1 \\ x_2 \\ \vdots \\ x_n \end{bmatrix}$ is a feasible solution and it

has p positive components, i.e., $x = \begin{bmatrix} x_1 \\ x_2 \\ \vdots \\ x_p \\ 0 \\ 0 \\ 0 \end{bmatrix}$. That is, $x_i \geq 0$,

where $i = 1, \ldots, p$. Let $A = \begin{bmatrix} a_1 & a_2 & \ldots & a_p & \ldots & a_n \end{bmatrix}$, where a_i for all $i = 1, \ldots, n$, is the ith column of A. We have

$$Ax = b,$$

then

$$\begin{bmatrix} a_1 & a_2 & \ldots & a_p & \ldots & a_n \end{bmatrix} \begin{bmatrix} x_1 \\ x_2 \\ \vdots \\ x_p \\ 0 \\ 0 \\ 0 \end{bmatrix} = b.$$

That is,

$$x_1 a_1 + x_2 a_2 + \cdots + x_p a_p = b. \tag{5.3}$$

We consider two cases.

Case 1: If $a_1, a_2, \ldots, a_p$ are linearly independent, then $p \leq m$. If $p = m$, then the solution x is basic and the proof is done. If $p < m$ and since rank $A = m$, then the matrix B is square matrix of order $m \times m$ and therefore, $x = \begin{bmatrix} x_1 \\ x_2 \\ \vdots \\ x_p \\ 0 \\ 0 \\ 0 \end{bmatrix}$. That is,

$x_i > 0$ for all $i = 1, \ldots, p$, and $x_i = 0$ for all $i = p+1, \ldots, m$. Thus, x is a degenerate basic feasible solution corresponding to the basis B.

Case 2: If $a_1, a_2, \ldots, a_p$ are linearly dependent, then there exist α_i, $i = 1, \ldots, p$, not all zero, such that

$$\alpha_1 a_1 + \alpha_2 a_2 + \cdots + \alpha_p a_p = 0. \tag{5.4}$$

We can assume that there exists at least one α_i, that is positive. Note that if all α_i are nonpositive, then we can multiply (5.4) by (-1).

Multiplying (5.4) by ϵ, a real number,

$$\epsilon \alpha_1 a_1 + \epsilon \alpha_2 a_2 + \cdots + \epsilon \alpha_p a_p = 0. \tag{5.5}$$

Subtracting (5.5) from (5.3) to get

$$(x_1 - \epsilon \alpha_1)a_1 + (x_2 - \epsilon \alpha_2)a_2 + \cdots + (x_p - \epsilon \alpha_p)a_p = b.$$

Let $\alpha = \begin{bmatrix} \alpha_1 \\ \vdots \\ \alpha_p \\ 0 \\ 0 \\ 0 \end{bmatrix}$. Then, for any ϵ, we can write $A[x - \epsilon \alpha] = b$.

Let $\epsilon = \min\{\frac{x_i}{\alpha_i} : i = 1, \ldots, p, \ \alpha_i > 0\}$. Then, the first p components of $x - \epsilon \alpha$ are non-negative, and at least one of these components is zero. Therefore, we have a feasible solution with at most $p - 1$ positive components. We repeat this process until we get linearly independent columns of A, after that we get back to **case-1**. Therefore, Part 1 is done. We now prove Part 2.

2. Let the solution $x = \begin{bmatrix} x_1 \\ x_2 \\ \vdots \\ x_n \end{bmatrix}$ be an optimal feasible solution

and the first p components are nonzero.

We have two cases to consider. The first case is exactly the same as in Part 1. The second case follows the same arguments as in Part 1, but in addition, we must show that $x - \epsilon\alpha$ is the optimal for any ϵ. That is, we must show that $c^T\alpha = 0$.
Suppose that if possible $c^T\alpha \neq 0$. Note that for ϵ, sufficiently small,

$$(|\epsilon| \leq \min\{| \frac{x_i}{\alpha_i} |: i = 1, \ldots, p, \ \alpha_i \neq 0\}).$$

The vector $x - \epsilon\alpha$ is feasible. We choose ϵ such that

$$c^T x > c^T x - \epsilon c^T \alpha$$
$$= c^T(x - \epsilon\alpha).$$

It contradicts the optimality of x.

$\square$

Definition 5.5 (Extreme Points). A point x of a convex set Ω is called an extreme point of Ω if there are no two points x_1 and x_2 such that $\lambda x_1 + (1 - \lambda x_2) = x$ for some $\lambda \in [0, 1]$. It means that the extreme point is a point that does not lie strictly within the line segment connecting two other points of the set.

In other words, if x is an extreme point and $x = \lambda x_1 + (1 - \lambda)x_2$ for some $x_1, x_2 \in \Omega$, then $x_1 = x_2$.
For example,
Let the set R_0 be determined by

$$x_1 + x_2 + x_3 = 1,$$
$$x_1 - x_2 = 0,$$
$$x_1 \geq 0, x_2 \geq 0, x_3 \geq 0.$$

Note that R_0 is the line segment joining the points $x_1 = \begin{bmatrix} 1/2 \\ 1/2 \\ 0 \end{bmatrix}$ and

$x_2 = \begin{bmatrix} 0 \\ 0 \\ 1 \end{bmatrix}$. Both x_1 and x_2 are extreme points; however, only one of them, namely x_1, is nondegenerate.

Theorem 5.3. *Let Ω be the convex set consisting of all feasible solutions, that is, all n-vectors x satisfying*

$$Ax = b, \ x \geq 0, \tag{5.6}$$

where $A \in \mathbb{R}^{m \times n}$, $m < n$. Then, x is an extreme point of Ω if and only if x is a basic feasible solution to $Ax = b$, $x \geq 0$.

Proof. Suppose that $x = [x_1, x_2, \ldots, x_p, 0, 0, \ldots, 0]^T$ satisfies (5.6) and has p positive components. Without loss of generality, we can assume that first p components are positive and the remaining components are zero. We have

$$x_1 a_1 + x_2 a_2 + \cdots + x_p a_p = b, \tag{5.7}$$

Let α_i, $i = 1, \ldots, p$ be such that

$$\alpha_1 a_1 + \alpha_2 a_2 + \cdots + \alpha_p a_p = 0. \tag{5.8}$$

We show that each $\alpha_i = 0$, that is, we show that $a_1, a_2, \ldots, a_p$ are linearly independent. Let $\epsilon > 0$, then

$$\epsilon \alpha_1 a_1 + \epsilon \alpha_2 a_2 + \cdots + \epsilon \alpha_p a_p = 0. \tag{5.9}$$

Adding (5.7) and (5.9) and subtracting (5.9) from (5.7) to get

$$(x_1 + \epsilon \alpha_1) a_1 + (x_2 + \epsilon \alpha_2) a_2 + \cdots + (x_p + \epsilon \alpha_p) a_p = b,$$
$$(x_1 - \epsilon \alpha_1) a_1 + (x_2 - \epsilon \alpha_2) a_2 + \cdots + (x_p - \epsilon \alpha_p) a_p = b.$$

Since each $x_i > 0$, $\epsilon > 0$ can be chosen such that $x_i + \epsilon \alpha_i \geq 0$ and $x_i - \epsilon \alpha_i \geq 0$. We have

$$(|\epsilon| \leq \min\{| \frac{x_i}{\alpha_i} | : i = 1, \ldots, p, \ \alpha_i \neq 0\}).$$

For such a choice of ϵ, the vectors

$$z_1 = \begin{bmatrix} x_1 + \epsilon\alpha_1 \\ x_2 + \epsilon\alpha_2 \\ \vdots \\ x_p + \epsilon\alpha_p \\ 0 \\ 0 \\ 0 \end{bmatrix} \quad \text{and } z_2 = \begin{bmatrix} x_1 - \epsilon\alpha_1 \\ x_2 - \epsilon\alpha_2 \\ \vdots \\ x_p - \epsilon\alpha_p \\ 0 \\ 0 \\ 0 \end{bmatrix}, \quad \text{where } z_1, z_2 \in \Omega.$$

Note that $x = \frac{1}{2}z_1 + \frac{1}{2}z_2$, but x is extreme. Therefore, $z_1 = z_2$. That is $\alpha_i = 0$, $i = 1, \ldots, p$ and $\alpha_1, \alpha_2, \ldots, \alpha_p$ are linearly inde-

pendent. Thus, $x = \begin{bmatrix} x_1 \\ x_2 \\ \vdots \\ x_p \\ 0 \\ 0 \\ 0 \end{bmatrix}$ is a basic feasible solution to $Ax = b$,

where $x \geq 0$.

Conversely, let $x \in \Omega$ be a basic feasible solution. Let $y, z \in \Omega$ be such that $x = \alpha y + (1 - \alpha)z$ for some $\alpha \in (0, 1)$.
We need to show that x is extreme, i.e., equivalent to show that $y = z$.
Since $y, z \geq 0$ and the last $n - m$ components of x are zero. Therefore, the last $n - m$ components of y and z are zero as well.
Furthermore, since

$$Ay = Az = b,$$

then

$$y_1 a_1 + \cdots + y_m a_m = b, \tag{5.10}$$
$$z_1 a_1 + \cdots + z_m a_m = b. \tag{5.11}$$

Subtracting (5.11) from (5.10),

$$(y_1 - z_1)a_1 + \cdots + (y_m - z_m)a_m = 0.$$

Since the columns $a_1, a_2, \ldots, a_m$ are linearly independent, then $y_1 - z_1 = 0$, $y_2 - z_2 = 0, \ldots, y_m - z_m = 0$. That is,

$$y_i = z_i, \ i = 1, \ldots, m.$$

Thus, x is an extreme point of Ω. □

5.4 Simplex Algorithm

The aim of the simplex algorithm is to move from one basic feasible solution to another until an optimal basic feasible solution is found and the value of objective function continually decreases until a minimum is reached. A basic feasible solution is optimal if and only if the corresponding reduced cost coefficients are all non-negative.

Theorem 5.4. *A basic feasible solution is optimal if and only if the corresponding reduced cost coefficients are all non-negative.*

Note: The proof is beyond the scope of this book.

At this point, we have the following steps for the simplex algorithm.

Algorithm

1. Form an augmented simplex tableau corresponding to a starting basic feasible solution of the standard form of linear programming problem.

2. Calculate the reduced cost coefficients corresponding to the nonbasic variables.

3. If $r_j \geq 0$ for all j, then the current basic feasible solution is optimal and stop; otherwise, go to step 4.

4. Select a q such that $r_q < 0$ (i.e., the q corresponding to the most negative r_q).

5. If no $y_{iq} > 0$, then the problem is unbounded and stop; otherwise, go to step 6.

6. Calculate $p = \arg\min_i \left\{ \frac{y_{i0}}{y_{iq}} : y_{iq} > 0 \right\}$. If more than one index i minimizes $\frac{y_{i0}}{y_{iq}}$, then let us choose p.

7. Update the augmented simplex tableau by pivoting about the $(p, q)^{\text{th}}$ element.

8. Go to step 2.

The simplex method can be examined by selecting pivot column q and row p. The pivot column is q with the most negative element in the bottom row of an augmented matrix. The tableau is called an optimal in step 3 if all entries are non-negative in the last row of the tableau. In the simplex method, we want to move from one basic feasible solution to another until an optimal basic feasible solution is found. To improve the tableau, pivot row p is calculated by attaining minimum ratio in step 6 that is $\min_i \left\{ \frac{y_{i0}}{y_{iq}} \right\}$ where the entries $y_{i0}, i = 1, 2, \ldots, m$ are found in the last column of the canonical augmented matrix and y_{iq} are the entries of pivot column q. We use elementary row operations to get zero entries at q^{th} column except the $(p,q)^{\text{th}}$ entry which is made to be unity. The rest of the entries will remain the same as that of old canonical augmented matrix. If all entries y_{iq} are either zero or negative, then we cannot move to a new augmented matrix and the linear programming problem will be unbounded.

Example 5.3. Solve linear programming problem.

$$
\begin{aligned}
\text{maximize} \quad & 4x_1 + 5x_2 \\
\text{subject to} \quad & 2x_1 + x_2 \le 9, \\
& x_1 \quad\quad \le 4, \\
& \quad\quad x_2 \le 3, \\
& x_1, \quad x_2 \ge 0.
\end{aligned}
$$

The first step is to convert the problem in standard form:

$$
\begin{aligned}
\text{minimize} \quad & -4x_1 - 5x_2 \\
\text{subject to} \quad & 2x_1 + x_2 + x_3 && = 9, \\
& x_1 && +x_4 && = 4, \\
& x_2 && +x_5 = 3, \\
& x_1, \quad x_2, \quad x_3, \ x_4, \ x_5, \geq 0.
\end{aligned}
$$

The augmented matrix of this problem is

	a_1	a_2	a_3	a_4	a_5	b
a_3	2	1	1	0	0	9
a_4	1	0	0	1	0	4
a_5	0	1	0	0	1	3

Note that basis matrix is $B = \begin{bmatrix} a_3 & a_4 & a_5 \end{bmatrix}$. Therefore, $x_B = \begin{bmatrix} x_3 \\ x_4 \\ x_5 \end{bmatrix}$
and since $Bx_B = b$, so we get $x_B = \begin{bmatrix} 9 \\ 4 \\ 3 \end{bmatrix}$. Therefore, the starting

basic feasible solution to the problem is $x = \begin{bmatrix} 0 \\ 0 \\ 9 \\ 4 \\ 3 \end{bmatrix}$. We compute the

reduced cost coefficients corresponding to the nonbasic variables x_1 and x_2.

$$
\begin{aligned}
r_1 &= c_1 - \left(c_3 y_{11} + c_4 y_{21} + c_5 y_{31} \right) \\
&= -4 - (0 \times 2 + 0 \times 1 + 0 \times 0) \\
&= -4,
\end{aligned}
$$

and

$$
\begin{aligned}
r_2 &= c_2 - \left(c_3 y_{12} + c_4 y_{22} + c_5 y_{32} \right) \\
&= -5 - (0 \times 1 + 0 \times 0 + 0 \times 1) \\
&= -5.
\end{aligned}
$$

We choose the most negative cost coefficient value r_j and bring the corresponding columns into the basis. We see that $r_2 = -5$ is the most negative value. Therefore, a_2 enters into basis.

	a_1	a_2	a_3	a_4	a_5	b
a_3	2	1	1	0	0	9
a_4	1	0	0	1	0	4
a_5	0	1	0	0	1	3

We compute p= arg min$\left\{\frac{9}{1}, \frac{4}{0}, \frac{3}{1}\right\} = 3^{\text{rd}}$ row. We choose the minimum ratio and we see that the third row is chosen as pivot row.

	a_1	a_2	a_3	a_4	a_5	b
a_3	2	1	1	0	0	9
a_4	1	0	0	1	0	4
a_2	0	1	0	0	1	3

We now update the canonical augmented matrix by pivoting at $(3,2)^{\text{th}}$ entry with the help of row operations:

	a_1	a_2	a_3	a_4	a_5	b
a_3	2	1	1	0	0	9
a_4	1	0	0	1	0	4
a_2	0	1	0	0	1	3

$R_1 \to R_1 - R_3$

	a_1	a_2	a_3	a_4	a_5	b
a_3	2	0	1	0	-1	6
a_4	1	0	0	1	0	4
a_2	0	1	0	0	1	3

$R_1 \leftrightarrow R_3$

	a_1	a_2	a_3	a_4	a_5	b
a_2	0	1	0	0	1	3
a_4	1	0	0	1	0	4
a_3	2	0	1	0	-1	6

$R_2 \leftrightarrow R_3$

	a_1	a_2	a_3	a_4	a_5	b
a_2	0	1	0	0	1	3
a_3	2	0	1	0	-1	6
a_4	1	0	0	1	0	4

We get basis $B = \begin{bmatrix} a_2 & a_3 & a_4 \end{bmatrix}$, $x_B = \begin{bmatrix} x_2 \\ x_3 \\ x_4 \end{bmatrix}$. Since $Bx_B = b$,

therefore we get $x_B = \begin{bmatrix} 3 \\ 6 \\ 4 \end{bmatrix}$ and hence $x = \begin{bmatrix} 0 \\ 3 \\ 6 \\ 4 \\ 0 \end{bmatrix}$.

We again compute

$$r_1 = c_1 - (c_2 y_{11} + c_3 y_{21} + c_4 y_{31})$$
$$= -4 - (-5 \times 0 + 0 \times 2 + 0 \times 1)$$
$$= -4,$$

and

$$r_5 = c_5 - (c_2 y_{15} + c_3 y_{25} + c_4 y_{35})$$
$$= 0 - (-5 \times 1 + 0 \times -1 + 0 \times 0)$$
$$= 5.$$

Since r_1 is the only negative reduced cost, therefore we choose first column a_1 to enter into basis.

	a_1	a_2	a_3	a_4	a_5	b
a_2	0	1	0	0	1	3
a_3	2	0	1	0	−1	6
a_4	1	0	0	1	0	4

We find p $= \arg\min\{\frac{3}{0}, \frac{6}{2}, \frac{4}{1}\} = 2^{\text{nd}}$ row, therefore row 2 should be the pivot row.

	a_1	a_2	a_3	a_4	a_5	b
a_2	0	1	0	0	1	3
a_1	2	0	1	0	−1	6
a_4	1	0	0	1	0	4

We again perform elementary row operations by pivoting at $(2, 1)^{\text{th}}$ entry.

	a_1	a_2	a_3	a_4	a_5	b
a_2	0	1	0	0	1	3
a_1	2	0	1	0	−1	6
a_4	1	0	0	1	0	4

$R_2 \to \frac{1}{2}R_2$

	a_1	a_2	a_3	a_4	a_5	b
a_2	0	1	0	0	1	3
a_1	1	0	1/2	0	−1/2	3
a_4	1	0	0	1	0	4

$R_3 \to R_3 - R_2$

	a_1	a_2	a_3	a_4	a_5	b
a_2	0	1	0	0	1	3
a_1	$\boxed{1}$	0	$1/2$	0	$-1/2$	3
a_4	0	0	$-1/2$	1	$1/2$	1

$R_1 \leftrightarrow R_2$

	a_1	a_2	a_3	a_4	a_5	b
a_1	1	0	$1/2$	0	$-1/2$	3
a_2	0	1	0	0	1	3
a_4	0	0	$-1/2$	1	$1/2$	1

Thus, basis $B = [a_1 \quad a_2 \quad a_4]$, $x_B = \begin{bmatrix} x_1 \\ x_2 \\ x_4 \end{bmatrix}$, so $Bx_B = b$ implies

$$x_B = \begin{bmatrix} 3 \\ 3 \\ 1 \end{bmatrix} \text{ and } x = \begin{bmatrix} 3 \\ 3 \\ 0 \\ 1 \\ 0 \end{bmatrix}.$$

We find

$$r_3 = c_3 - (c_1 y_{13} + c_2 y_{23} + c_4 y_{33})$$
$$= 0 - (-4 \times \frac{1}{2} + (-5) \times 0 + 0 \times -\frac{1}{2})$$
$$= 2,$$

and

$$r_5 = c_5 - (c_1 y_{15} + c_2 y_{25} + c_4 y_{35})$$
$$= 0 - (-4 \times (-\frac{1}{2}) + (-5) \times 1 + 0 \times \frac{1}{2})$$
$$= 3.$$

Note that all the reduced costs are non-negative. Therefore, the current basic feasible solution is the optimal solution. Thus, $x_1 = 3$ and $x_2 = 3$, and the value of the objective function is

$$4x_1 + 5x_2 = 4 \times 3 + 5 \times 3 = 27.$$

Example 5.4. Solve the linear programming problem using the simplex method.

$$\begin{aligned} \text{maximize} \quad & 5x_1 + 6x_2 \\ \text{subject to} \quad & 3x_1 + 4x_2 \leq 18, \\ & 2x_1 + x_2 \leq 7, \\ & x_1, \quad x_2 \geq 0. \end{aligned}$$

We first transform the problem into standard form:

$$\begin{aligned} \text{minimize} \quad & -5x_1 - 6x_2 \\ \text{subject to} \quad & 3x_1 + 4x_2 + x_3 = 18, \\ & 2x_1 + x_2 + x_4 = 7, \\ & x_1, \quad x_2, \quad x_3, \quad x_4 \geq 0. \end{aligned}$$

We construct the augmented matrix for the linear programming problem:

	a_1	a_2	a_3	a_4	b
a_3	3	4	1	0	18
a_4	2	1	0	1	7
c^T	-5	-6	0	0	0

We see that the basis matrix is $B = \begin{bmatrix} a_3 & a_4 \end{bmatrix}$, therefore basic vector is $x_B = \begin{bmatrix} x_3 \\ x_4 \end{bmatrix}$ and since $Bx_B = b$, so we get $x_B = \begin{bmatrix} 18 \\ 7 \end{bmatrix}$. Therefore, the starting basic feasible solution is $x = \begin{bmatrix} 0 \\ 0 \\ 18 \\ 7 \end{bmatrix}$.

We choose the most negative element in the last row of the tableau corresponding to nonbasic variables. Since $r_2 = -6$ is most negative element, we get the second column as a pivot column, i.e., q=2 and bring a_2 into basis. We compute ratios

$$\frac{y_{10}}{y_{12}} = \frac{18}{4} = \frac{9}{2}, \qquad \frac{y_{20}}{y_{22}} = \frac{7}{1} = 7.$$

We find the minimum value as $\frac{9}{2}$ which is for the first row, i.e., p=1. Thus, a_3 of basis matrix B is replaced by a_2.

We now perform elementary row operations to get zero entries in the second pivot column except $(p,q)^{th} = (1,2)^{th}$ place which is the unity. This is called update of tableau.

	a_1	a_2	a_3	a_4	b
a_2	3	4	1	0	18
a_4	2	1	0	1	7
c^T	-5	-6	0	0	0

$R_1 \to \frac{R_1}{4}$

	a_1	a_2	a_3	a_4	b
a_2	$3/4$	1	$1/4$	0	$9/2$
a_4	2	1	0	1	7
c^T	-5	-6	0	0	0

$R_2 \to R_2 - R_1$
$R_3 \to R_3 + 6R_1$

	a_1	a_2	a_3	a_4	b
a_2	$3/4$	1	$1/4$	0	$9/2$
a_4	$5/4$	0	$-1/4$	1	$5/2$
c^T	$-1/2$	0	$3/2$	0	27

We again check the reduced cost coefficient; we see that only $r_1 = -\frac{1}{2}$ is most negative. Therefore, q=1, i.e., we bring a_1 into basis. Since

$$\frac{y_{10}}{y_{11}} = 6, \qquad \frac{y_{20}}{y_{21}} = 2.$$

The minimum value is 2, therefore we choose p=2 as the pivot row and a_4 is replaced by a_1. Thus, we update the tableau by pivoting about $(p,q)^{th} = (2,1)^{th}$ element to obtain

	a_1	a_2	a_3	a_4	b
a_2	$3/4$	1	$1/4$	0	$9/2$
a_1	$5/4$	0	$-1/4$	1	$5/2$
c^T	$-1/2$	0	$3/2$	0	27

$R_2 \to \frac{4}{5} R_2$

	a_1	a_2	a_3	a_4	b
a_2	$3/4$	1	$1/4$	0	$9/2$
a_1	1	0	$-1/5$	$4/5$	2
c^T	$-1/2$	0	$3/2$	0	27

$R_1 \to R_1 - \frac{3}{4} R_2$
$R_3 \to R_3 + \frac{1}{2} R_2$

	a_1	a_2	a_3	a_4	b
a_2	0	1	$2/5$	$-3/5$	3
a_1	1	0	$-1/5$	$4/5$	2
c^T	0	0	$7/5$	$2/5$	28

$R_1 \leftrightarrow R_2$

	a_1	a_2	a_3	a_4	b
a_1	1	0	$-1/5$	$4/5$	2
a_2	0	1	$2/5$	$-3/5$	3
c^T	0	0	$7/5$	$2/5$	28

Since the last row of this tableau has no negative elements, we conclude that the basic feasible solution corresponding to this tableau is optimal. Thus,

$$x_1 = 2, \ x_2 = 3, \ x_3 = 0 \text{ and } x_4 = 0$$

is the solution to the original linear programming problem in standard form and the corresponding objective value is 28.

Example 5.5. Solve linear programming problem using the simplex method.

$$\begin{aligned}
\text{maximize} \quad & 3x_1 + 6x_2 + 2x_3 \\
\text{subject to} \quad & 3x_1 + 4x_2 + x_3 \leq 20, \\
& x_1 + 3x_2 + 2x_3 \leq 10, \\
& x_1, \quad x, \quad x_3 \geq 0.
\end{aligned}$$

We write the linear programming problem in standard form:

$$\begin{aligned}
\text{minimize} \quad & -3x_1 - 6x_2 - 2x_3 \\
\text{subject to} \quad & 3x_1 + 4x_2 + x_3 + x_4 = 20, \\
& x_1 + 3x_2 + 2x_3 + x_5 = 10, \\
& x_1, \quad x_2, \quad x_3, \quad x_4, \quad x_5 \geq 0.
\end{aligned}$$

We create the simplex tableau as

	a_1	a_2	a_3	a_4	a_5	b
a_4	3	4	1	1	0	20
a_5	1	3	2	0	1	10
c^T	-3	-6	-2	0	0	0

Note that in the above simplex tableau, the coefficient of variables of objective function is written in the last row of tableau. The current basis matrix is $B = \begin{bmatrix} a_4 & a_5 \end{bmatrix}$, therefore basic vector is $x_B = \begin{bmatrix} x_4 \\ x_5 \end{bmatrix}$ and since $Bx_B = b$, so we get $x_B = \begin{bmatrix} 20 \\ 10 \end{bmatrix}$.

The starting basic feasible solution is $x = \begin{bmatrix} x_1 \\ x_2 \\ x_3 \\ x_4 \\ x_5 \end{bmatrix} = \begin{bmatrix} 0 \\ 0 \\ 0 \\ 20 \\ 10 \end{bmatrix}$ and value of objective function is zero. We consider the negative coefficient corresponding to nonbasic variables in the simplex tableau. We have

$$r_1 = -3, r_2 = -6, r_3 = -2.$$

We choose the most negative element. Note that $r_2 = -6$ is the most negative. Therefore, we obtain q=2. That is, the second column is the pivot column.

	a_1	a_2	a_3	a_4	a_5	b
a_4	3	4	1	1	0	20
a_5	1	3	2	0	1	10
c^T	-3	-6	-2	0	0	0

We use the ratio test to choose the pivot row, i.e., p. For this, we use the sixth statement of the simplex algorithm.

$$p = \arg \min \left\{ \frac{y_{10}}{y_{12}}, \frac{y_{20}}{y_{22}} \right\} = \arg \min \left\{ \frac{20}{4}, \frac{10}{3} \right\} = 2^{\text{nd}} \text{ row.}$$

	a_1	a_2	a_3	a_4	a_5	b
a_4	3	4	1	1	0	20
a_5	1	3	2	0	1	10
c^T	-3	-6	-2	0	0	0

Thus, the second row is the pivot row. The intersecting box indicated is our pivot element and a_5 of basis matrix will be replaced by a_2 as a basic variable. We update the tableau by pivoting $(2, 2)^{\text{th}}$ element.

$$R_2 \rightarrow \tfrac{1}{3} R_2$$

	a_1	a_2	a_3	a_4	a_5	b
a_4	3	4	1	1	0	20
a_2	$1/3$	1	$2/3$	0	$1/3$	$10/3$
c^T	-3	-6	-2	0	0	0

$$R_1 \rightarrow R_1 - 4R_2$$
$$R_3 \rightarrow R_3 + 6R_2$$

	a_1	a_2	a_3	a_4	a_5	b
a_4	$5/3$	0	$-5/3$	1	$-4/3$	$20/3$
a_2	$1/3$	1	$2/3$	0	$1/3$	$10/3$
c^T	-1	0	2	0	2	20

We choose the most negative. Only $r_1 = -1$ is the most negative. Thus, the first column is the pivot column and bring a_1 into basis. To determine which element in the second column is the appropriate pivot, we compute the two ratios:

$$\mathrm{p} = \arg \min\left\{\frac{y_{10}}{y_{11}}, \frac{y_{20}}{y_{21}}\right\} = \left\{\frac{20}{5}, 10\right\} = 1^{\text{st}} \text{ row.}$$

This gives the first row as the pivot row.

	a_1	a_2	a_3	a_4	a_5	b
a_1	$5/3$	0	$-5/3$	1	$-4/3$	$20/3$
a_2	$1/3$	1	$2/3$	0	$1/3$	$10/3$
c^T	-1	0	2	0	2	20

We apply row operations to pivot the $(1,1)^{\text{th}}$ element for updating the tableau.

$R_1 \rightarrow \frac{3}{5}R_1$

	a_1	a_2	a_3	a_4	a_5	b
a_1	1	0	-1	$3/5$	$-4/5$	4
a_2	$1/3$	1	$2/3$	0	$1/3$	$10/3$
c^T	-1	0	2	0	2	20

$R_2 \rightarrow R_2 - \frac{1}{3}R_1$
$R_3 \rightarrow R_3 + R_1$

	a_1	a_2	a_3	a_4	a_5	b
a_1	1	0	-1	$3/5$	$-4/5$	4
a_2	0	1	1	$-1/5$	$3/5$	2
c^T	0	0	1	$3/5$	$6/5$	24

Since the last row has no negative elements, we conclude that the solution corresponding to the above tableau is optimal. Thus, $x = \begin{bmatrix} x_1 \\ x_2 \\ x_3 \\ x_4 \end{bmatrix} = \begin{bmatrix} 4 \\ 2 \\ 0 \\ 0 \end{bmatrix}$ and the optimal solution of the maximum objective function is 24.

We can solve linear programming problems with the help of MATLAB. To choose the pivot column from the feasible tableau, we use MATLAB function **pivotcolumn.m** given in the following Code 5.1:

Code 5.1: pivotcolumn.m

```
function [A,q]=pivotcolumn(A,v)
%input :augmented matrix A, nonbasic
%variables v
%output:pivot column q, augmented matrix A
    [m,~]=size(A);
    min=0;
    q=0;
    for J=1:v
        if A(m,J)<0
            if A(m,J)<min
            min=A(m,J);
            q=J;
        end
    end
    end
    return
```

MATLAB function **pivotrow.m** given in Code 5.2 is written to choose the pivot row. This function also checks whether the linear programming problem is unbounded or not.

Code 5.2: pivotrow.m

```
function [A,p,e,B]=pivotrow(A,q,B)
%input : augmented matrix A, pivot column q,
%         basis vector B
%output: pivot row p, pivot element e
    [m,n]=size(A);
    min=Inf;
    p=0;
    for k=1:m
        if A(k,q)>0
            row=A(k,n)/A(k,q);
            if row<min
                min=row;
                p=k;
            end
        end
    end
    if p==0
        disp('unbounded');
        e=0;
    else
        e=A(p,q);
        B(p)=q;
    end
    return
```

The MATLAB function `simplex.m` is given in Code 5.3 to make zero entries in q^{th} column except $(p,q)^{th}$ place which is unity.

Code 5.3: simplex.m

```
function A=simplex(A,p,q)
%input:  augmented matrix A, pivot row p,
%         pivot column q
%output: augmented matrix A
    [m,n]=size(A);
    format rat
    C=zeros(m,n);
```

```
   J=(1:n);
   C(p,J)=sym(A(p,J)/A(p,q));
    for  I=1:m
        if  I~=p
C(I,J)=sym(A(I,J))−sym(A(I,q))*C(p,J);
        end
    end
    A=C;
   return
```

We now see how the simplex algorithm works in MATLAB to move from a feasible tableau to an optimal tableau, one pivot at a time, by means of the following examples:

Example 5.6. Solve the linear programming problem using the simplex method.

$$\text{maximize} \quad 7x_1 + 6x_2$$
$$\text{subject to} \quad 2x_1 + x_2 \leq 3,$$
$$x_1 + 4x_2 \leq 4,$$
$$x_1, \quad x_2 \geq 0.$$

We first transform the problem into standard form. To do this, we multiply the objective function by -1 to change the maximization to minimization. We introduce two non-negative slack variables x_3 and x_4. Then, the original problem can be written as

$$\text{minimize} \quad -7x_1 - 6x_2$$
$$\text{subject to} \quad 2x_1 + x_2 + x_3 \quad = 3,$$
$$x_1 + 4x_2 \quad +x_4 = 4,$$
$$x_1, \quad x_2, \quad x_3, \quad x_4 \geq 0.$$

We construct the tableau for the problem:

$$
\begin{bmatrix}
a_1 & a_2 & a_3 & a_4 & b \\
2 & 1 & 1 & 0 & 3 \\
1 & 4 & 0 & 1 & 4 \\
-7 & -6 & 0 & 0 & 0
\end{bmatrix}
$$

Note that the basis matrix is $B = [a_3 \quad a_4], x_B = \begin{bmatrix} x_3 \\ x_4 \end{bmatrix}$, and

$Bx_B = b$ implies $x_B = \begin{bmatrix} 3 \\ 4 \end{bmatrix}$. Therefore, the starting basic feasi-

ble solution is $x = \begin{bmatrix} 0 \\ 0 \\ 3 \\ 4 \end{bmatrix}$. The last row contains the reduced cost

coefficients.

We use MATLAB function `pivotcolumn.m` to choose the pivot column corresponding to the most negative value. Note that non-basic variables are x_1 and x_2. Therefore, we take v=2.

```
>>[A,q]=pivotcolumn(A,v)
```

	a_1	a_2	a_3	a_4	b
a_3	2	1	1	0	3
a_4	1	4	0	1	4
c^T	-7	-6	0	0	0

q=1.

We compute p = arg min$\left\{ \dfrac{y_{10}}{y_{11}}, \dfrac{y_{20}}{y_{21}} \right\} = \left\{ \dfrac{3}{2}, \dfrac{4}{1} \right\} = 1^{\text{st}}$ row.

```
>>B=[3 4]
>>[A,p,e,B]=pivotrow(A,q,B)
```

	a_1	a_2	a_3	a_4	b
a_3	2	1	1	0	3
a_4	1	4	0	1	4
c^T	-7	-6	0	0	0

p = 1   e = 2   B = $\begin{bmatrix} 1 & 4 \end{bmatrix}$.

$$
\begin{array}{c c c c c c}
 & a_1 & a_2 & a_3 & a_4 & b \\
a_1 & \boxed{2} & 1 & 1 & 0 & 3 \\
a_4 & 1 & 4 & 0 & 1 & 4 \\
c^T & -7 & -6 & 0 & 0 & 0
\end{array}
$$

$$R_1 \to \tfrac{1}{2}R_1$$

```
>>A=identityop(A,p,½)
```

$$
\begin{array}{c c c c c c}
 & a_1 & a_2 & a_3 & a_4 & b \\
a_1 & \boxed{1} & 1/2 & 1/2 & 0 & 3/2 \\
a_4 & 1 & 4 & 0 & 1 & 4 \\
C^T & -7 & -6 & 0 & 0 & 0
\end{array}
$$

$$R_2 \to R_2 - R_1$$

```
>>A=eliminationop(A,2,p,-1)
```

$$
\begin{array}{c c c c c c}
 & a_1 & a_2 & a_3 & a_4 & b \\
a_1 & \boxed{1} & 1/2 & 1/2 & 0 & 3/2 \\
a_4 & 0 & 7/2 & -1/2 & 1 & 5/2 \\
c^T & -7 & -6 & 0 & 0 & 0
\end{array}
$$

$$R_3 \to R_3 + 7R_1$$

```
>>A=eliminationop(A,3,p,7)
```

$$
\begin{array}{c c c c c c}
 & a_1 & a_2 & a_3 & a_4 & b \\
a_1 & 1 & 1/2 & 1/2 & 0 & 3/2 \\
a_4 & 0 & 7/2 & -1/2 & 1 & 5/2 \\
c^T & 0 & -5/2 & 7/2 & 0 & 21/2
\end{array}
$$

```
>>[A,q]=pivotcolumn(A,v)
```

	a_1	a_2	a_3	a_4	b
a_1	1	$1/2$	$1/2$	0	$3/2$
a_4	0	$7/2$	$-1/2$	1	$5/2$
c^T	0	$-5/2$	$7/2$	0	$21/2$

q = 2.

```
>>B=[1  4]
>>[A,p,e,B]=pivotrow(A,q,B)
```

	a_1	a_2	a_3	a_4	b
a_1	1	$1/2$	$1/2$	0	$3/2$
a_4	0	$7/2$	$-1/2$	1	$5/2$
c^T	0	$-5/2$	$7/2$	0	$21/2$

p = 2   e = $7/2$ B = $\begin{bmatrix} 1 & 2 \end{bmatrix}$.

	a_1	a_2	a_3	a_4	b
a_1	1	$1/2$	$1/2$	0	$3/2$
a_2	0	$7/2$	$-1/2$	1	$5/2$
c^T	0	$-5/2$	$7/2$	0	$21/2$

$$R_2 \to \tfrac{2}{7}R_2$$

```
>>A=identityop(A,p,1/e)
```

	a_1	a_2	a_3	a_4	b
a_1	1	$1/2$	$1/2$	0	$3/2$
a_2	0	1	$-1/7$	$2/7$	$5/7$
c^T	0	$-5/2$	$7/2$	0	$21/2$

$$R_3 \to R_3 + 5/2 R_2$$

```
>>A=eliminationop(A,3,p,5/2)
```

$$
\begin{array}{c c c c c c}
 & a_1 & a_2 & a_3 & a_4 & b \\
a_1 & 1 & 1/2 & 1/2 & 0 & 3/2 \\
a_2 & 0 & \boxed{1} & -1/7 & 2/7 & 5/7 \\
c^T & 0 & 0 & 22/7 & 5/7 & 86/7 \\
\end{array}
$$

$$R_1 \rightarrow R_1 - 1/2 R_2$$

```
>>A=eliminationop(A,1,p,-1/2)
```

$$
\begin{array}{c c c c c c}
 & a_1 & a_2 & a_3 & a_4 & b \\
a_1 & 1 & 0 & 4/7 & -1/7 & 8/7 \\
a_2 & 0 & 1 & -1/7 & 2/7 & 5/7 \\
c^T & 0 & 0 & 22/7 & 5/7 & 86/7 \\
\end{array}
$$

Since each element of the last row is non-negative, the present basic feasible solution is the optimal solution. Therefore, $x_1 = \frac{8}{7}, x_2 = \frac{5}{7}$ and the optimal value of objective function is $\frac{86}{7}$.

Example 5.7. Solve the linear programming problem.

$$
\begin{aligned}
\text{maximize} \quad & 2x_1 + x_2 \\
\text{subject to} \quad & x_1 \leq 5, \\
& x_2 \leq 7, \\
& x_1 + x_2 \leq 9, \\
& x_1, \quad x_2 \geq 0.
\end{aligned}
$$

The standard form of linear programming problem:

$$
\begin{aligned}
\text{minimize} \quad & -2x_1 - x_2 \\
\text{subject to} \quad & x_1 \quad\quad +x_3 \quad\quad\quad = 5, \\
& x_2 \quad\quad +x_4 \quad\quad = 7, \\
& x_1 + x_2 \quad\quad\quad +x_5 = 9, \\
& x_1, \quad x_2, \quad x_3, \quad x_4, \quad x_5 \geq 0.
\end{aligned}
$$

We construct the augmented matrix. It is represented by A.

$$
\begin{array}{c c c c c c c}
 & a_1 & a_2 & a_3 & a_4 & a_5 & b \\
a_3 & 1 & 0 & 1 & 0 & 0 & 5 \\
a_4 & 0 & 1 & 0 & 1 & 0 & 7 \\
a_5 & 1 & 1 & 0 & 0 & 1 & 9 \\
c^T & -2 & -1 & 0 & 0 & 0 & 0
\end{array}
$$

We have $B = \begin{bmatrix} a_3 & a_4 & a_5 \end{bmatrix}$, $x_B = \begin{bmatrix} x_3 \\ x_4 \\ x_5 \end{bmatrix}$ and $Bx_B = b$ implies

$x_B = \begin{bmatrix} 5 \\ 7 \\ 9 \end{bmatrix}$. Thus, the starting basic feasible solution is $x = \begin{bmatrix} 0 \\ 0 \\ 5 \\ 7 \\ 9 \end{bmatrix}$.

```
>>v=2
>>[A,q]=pivotcolumn(A,v)
```

$$
\begin{array}{c|c c c c c c}
 & a_1 & a_2 & a_3 & a_4 & a_5 & b \\
a_3 & 1 & 0 & 1 & 0 & 0 & 5 \\
a_4 & 0 & 1 & 0 & 1 & 0 & 7 \\
a_5 & 1 & 1 & 0 & 0 & 1 & 9 \\
c^T & -2 & -1 & 0 & 0 & 0 & 0
\end{array}
$$

q = 1.

We compute for pivot row p $= \arg \min \left\{ \frac{5}{1}, \frac{7}{0}, \frac{9}{1} \right\} = 1^{\text{st}}$ row.

```
>>B= [3  4  5]
>>[A,p,e,B]=pivotrow(A,q,B)
```

$$
\begin{array}{c|c c c c c c}
 & a_1 & a_2 & a_3 & a_4 & a_5 & b \\
a_3 & 1 & 0 & 1 & 0 & 0 & 5 \\
a_4 & 0 & 1 & 0 & 1 & 0 & 7 \\
a_5 & 1 & 1 & 0 & 0 & 1 & 9 \\
c^T & -2 & -1 & 0 & 0 & 0 & 0
\end{array}
$$

p = 1   e = 1   B = $\begin{bmatrix} 1 & 4 & 5 \end{bmatrix}$.

	a_1	a_2	a_3	a_4	a_5	b
a_1	1	0	1	0	0	5
a_4	0	1	0	1	0	7
a_5	1	1	0	0	1	9
c^T	-2	-1	0	0	0	0

$$R_3 \to R_3 - R_1$$

```
>>A=eliminationop(A,3,p,-1)
```

	a_1	a_2	a_3	a_4	a_5	b
a_1	1	0	1	0	0	5
a_4	0	1	0	1	0	7
a_5	0	1	-1	0	1	4
c^T	-2	-1	0	0	0	0

$$R_4 \to R_4 + 2R_1$$

```
>>A=eliminationop(A,4,p,2)
```

	a_1	a_2	a_3	a_4	a_5	b
a_1	1	0	1	0	0	5
a_4	0	1	0	1	0	7
a_5	0	1	-1	0	1	4
c^T	0	-1	2	0	0	10

```
>>[A,q]=pivotcolumn(A,v)
```

	a_1	a_2	a_3	a_4	a_5	b
a_1	1	0	1	0	0	5
a_4	0	1	0	1	0	7
a_5	0	1	-1	0	1	4
c^T	0	-1	2	0	0	10

```
q = 2.
```

Again, we compute pivot row p $= \arg \min\left\{\frac{\not{5}}{\not{0}}, \frac{7}{1}, \frac{4}{1}\right\} = 3^{\text{rd}}$ row.

```
>> B = [1  4  5]
>>[A,p,e,B]=pivotrow(A,q,B)
```

	a_1	a_2	a_3	a_4	a_5	b
a_1	1	0	1	0	0	5
a_4	0	1	0	1	0	7
a_5	0	1	-1	0	1	4
c^T	0	-1	2	0	0	10

p = 3   e = 1   B = $\begin{bmatrix} 1 & 4 & 2 \end{bmatrix}$.

	a_1	a_2	a_3	a_4	a_5	b
a_1	1	0	1	0	0	5
a_4	0	1	0	1	0	7
a_2	0	1	-1	0	1	4
c^T	0	-1	2	0	0	10

$$R_2 \to R_2 - R_3$$

We update the tableau using elementary row operations. We see that pivot element is already 1. Hence, we call only MATLAB function eliminationop.m

```
>>A=eliminationop(A,2,p,-1)
```

	a_1	a_2	a_3	a_4	a_5	b
a_1	1	0	1	0	0	5
a_4	0	0	1	1	-1	3
a_2	0	1	-1	0	1	4
c^T	0	-1	2	0	0	10

$$R_4 \to R_4 + R_3$$

```
>>A=eliminationop(A,4,p,1)
```

	a_1	a_2	a_3	a_4	a_5	b
a_1	1	0	1	0	0	5
a_4	0	0	1	1	−1	3
a_2	0	1	−1	0	1	4
c^T	0	0	1	0	1	14

$$R_2 \leftrightarrow R_3$$

```
>>A=exchangeop(A,2,3)
```

	a_1	a_2	a_3	a_4	a_5	b
a_1	1	0	1	0	0	5
a_2	0	1	−1	0	1	4
a_4	0	0	1	1	−1	3
c^T	0	0	1	0	1	14

Since the last row has no negative elements corresponding to non-basic variables, we conclude that the solution corresponding to the above tableau is optimal. Thus, we get $B = \begin{bmatrix} a_1 & a_2 & a_4 \end{bmatrix}$, $x_B = \begin{bmatrix} x_1 \\ x_2 \\ x_4 \end{bmatrix}$. Therefore, an optimal solution is $x = \begin{bmatrix} 5 \\ 4 \\ 0 \\ 3 \\ 0 \end{bmatrix}$ and the maximum value of an objective function is 14.

Example 5.8. Solve the linear programming problem using the simplex method.

$$
\begin{aligned}
\text{minimize} \quad & -2x_1 - x_2 \\
\text{subject to} \quad & -x_1 - x_2 \leq 1, \\
& x_1 - 2x_2 \leq 2, \\
& x_1, \quad x_2, \geq 0.
\end{aligned}
$$

We write the linear programming problem in standard form:

$$\begin{array}{ll} \text{minimize} & -2x_1 - x_2 \\ \text{subject to} & -x_1 - x_2 + x_3 = 1, \\ & x_1 - 2x_2 + x_4 = 2, \\ & x_1, \quad x_2, \quad x_3, \quad x_4, \geq 0. \end{array}$$

The problem can be written in an augmented matrix form:

	a_1	a_2	a_3	a_4	b
a_3	-1	-1	1	0	1
a_4	1	-2	0	1	2
c^T	-2	-1	0	0	0

Since the basis matrix is $B = \begin{bmatrix} a_3 & a_4 \end{bmatrix}$, $x_B = \begin{bmatrix} x_3 \\ x_4 \end{bmatrix}$ and $Bx_B = b$

implies $x_B = \begin{bmatrix} 1 \\ 2 \end{bmatrix}$. Therefore, the starting basic feasible solution

is $x = \begin{bmatrix} 0 \\ 0 \\ 1 \\ 2 \end{bmatrix}$. Note that we have two nonbasic variables x_1 and x_2.

Therefore, we take $v = 2$.

```
>>v=2
>>[A,q]=pivotcolumn[A,v]
```

	a_1	a_2	a_3	a_4	b
a_3	-1	-1	1	0	1
a_4	1	-2	0	1	2
c^T	-2	-1	0	0	0

q = 1.

We determine pivot row p = arg $\min\left\{ \cancel{\frac{1}{-1}}, \frac{2}{1} \right\}$ = 2nd row.

```
>> B = [3  4]
>>[A,p,e,B]=pivotrow(A,q,B)
```

$$
\begin{array}{c|ccccc}
 & a_1 & a_2 & a_3 & a_4 & b \\
\hline
a_3 & \boxed{-1} & -1 & 1 & 0 & 1 \\
a_4 & 1 & -2 & 0 & 1 & 2 \\
c^T & -2 & -1 & 0 & 0 & 0 \\
\end{array}
$$

p = 2   e = 1   B $=\begin{bmatrix} 3 & 1 \end{bmatrix}$.

$$
\begin{array}{c|ccccc}
 & a_1 & a_2 & a_3 & a_4 & b \\
a_3 & -1 & -1 & 1 & 0 & 1 \\
a_1 & \boxed{1} & -2 & 0 & 1 & 2 \\
c^T & -2 & -1 & 0 & 0 & 0 \\
\end{array}
$$

$$R_1 \rightarrow R_1 + R_2$$

```
>>A=eliminationop(A,1,p,1)
```

$$
\begin{array}{c|ccccc}
 & a_1 & a_2 & a_3 & a_4 & b \\
a_3 & 0 & -3 & 1 & 1 & 3 \\
a_1 & \boxed{1} & -2 & 0 & 1 & 2 \\
c^T & -2 & -1 & 0 & 0 & 0 \\
\end{array}
$$

$$R_3 \rightarrow R_3 + 2R_2$$

```
>>A=eliminationop(A,3,p,2)
```

$$
\begin{array}{c|ccccc}
 & a_1 & a_2 & a_3 & a_4 & b \\
a_3 & 0 & -3 & 1 & 1 & 3 \\
a_1 & 1 & -2 & 0 & 1 & 2 \\
c^T & 0 & -5 & 0 & 2 & 4 \\
\end{array}
$$

```
>> [A,q] = pivotcolumn(A,v)
>> [A,p,e,B] = pivotrow(A,q,B)
```

$$
\begin{array}{c|ccccc}
 & a_1 & a_2 & a_3 & a_4 & b \\
a_3 & 0 & \boxed{-3} & 1 & 1 & 3 \\
a_1 & 1 & -2 & 0 & 1 & 2 \\
c^T & 0 & -5 & 0 & 2 & 4 \\
\end{array}
$$

q = 2   unbounded.

We have p $=$ arg min$\left\{\cancel{\dfrac{3}{4}}, \cancel{\dfrac{2}{2}}\right\}$=no row. There is no such p. Therefore, the problem is unbounded.

Example 5.9. Solve the linear programming problem using the simplex method.

$$\begin{aligned}
\text{maximize} \quad & 45x_1 + 80x_2 \\
\text{subject to} \quad & 5x_1 + 20x_2 \le 400, \\
& 10x_1 + 15x_2 \le 450, \\
& x_1, \quad x_2 \ge 0.
\end{aligned}$$

The standard form of linear programming problem:

$$\begin{aligned}
\text{minimize} \quad & -45x_1 - 80x_2 \\
\text{subject to} \quad & 5x_1 + 20x_2 + x_3 \quad\quad = 400, \\
& 10x_1 + 15x_2 \quad\quad +x_4 = 450, \\
& x_1, \quad x_2, \quad x_3, \quad x_4 \ge 0.
\end{aligned}$$

The augmented matrix is written as

	a_1	a_2	a_3	a_4	b
a_3	5	20	1	0	400
a_4	10	15	0	1	450
c^T	-45	-80	0	0	0

Since the basis matrix is B=$\begin{bmatrix} a_3 & a_4 \end{bmatrix}$ and $x_B = \begin{bmatrix} x_3 \\ x_4 \end{bmatrix}$, therefore $Bx_B = b$ implies $x_B = \begin{bmatrix} 400 \\ 450 \end{bmatrix}$.

```
>> v = 2
>> [A,q] = pivotcolumn[A,v]
```

	a_1	a_2	a_3	a_4	b
a_3	5	20	1	0	400
a_4	10	15	0	1	450
c^T	-45	-80	0	0	0

q = 2.

The pivot row is $p = \arg\min\left\{\frac{400}{20}, \frac{450}{15}\right\} = \arg\min\{20, 30\} = 1^{\text{st}}$ row.

```
>> B = [3  4]
>> [A,p,e,B] = pivotrow(A,q,B)
```

	a_1	a_2	a_3	a_4	b
a_3	5	20	1	0	400
a_4	10	15	0	1	450
c^T	-45	-80	0	0	0

p = 1   e = 20   B = $\begin{bmatrix} 2 & 4 \end{bmatrix}$.

	a_1	a_2	a_3	a_4	b
a_2	5	20	1	0	400
a_4	10	15	0	1	450
c^T	-45	-80	0	0	0

$$R_1 \to \tfrac{1}{20}R_1$$

```
>> A = identityop(A,p,1/20)
```

	a_1	a_2	a_3	a_4	b
a_2	$1/4$	1	$1/20$	0	20
a_4	10	15	0	1	450
c^T	-45	-80	0	0	0

$$R_3 \to R_3 + 80R_1$$

```
>> A = eliminationop(A,3,p,80)
```

	a_1	a_2	a_3	a_4	b
a_2	$1/4$	1	$1/20$	0	20
a_4	10	15	0	1	450
c^T	-25	0	4	0	1600

$$R_2 \to R_2 - 15R_1$$

```
>> A = eliminationop(A,2,p,-15)
```

	a_1	a_2	a_3	a_4	b
a_2	$1/4$	1	$1/20$	0	20
a_4	$25/4$	0	$-3/4$	1	150
c^T	-25	0	4	0	1600

```
>>[A,q]=pivotcolumn(A,v)
```

	a_1	a_2	a_3	a_4	b
a_2	$1/4$	1	$1/20$	0	20
a_4	$25/4$	0	$-3/4$	1	150
c^T	-25	0	4	0	1600

q=1.

```
>>[A,p,e,B]=pivotrow(A,q,B)
```

	a_1	a_2	a_3	a_4	b
a_2	$1/4$	1	$1/20$	0	20
a_4	$25/4$	0	$-3/4$	1	150
c^T	-25	0	4	0	1600

p = 2   e = $25/4$ B = $\begin{bmatrix} 2 & 1 \end{bmatrix}$.

We now compute the pivot row $p = \arg\min\left\{\frac{20}{1/4}, \frac{150}{25/4}\right\}$ =$\arg\min\{80, 24\} = 2^{\text{nd}}$ row.

	a_1	a_2	a_3	a_4	b
a_2	$1/4$	1	$1/20$	0	20
a_1	$25/4$	0	$-3/4$	1	150
c^T	-25	0	4	0	1600

$$R_2 \to \tfrac{4}{25}R_2$$

```
>> A = identityop(A,p,4/25)
```

	a_1	a_2	a_3	a_4	b
a_2	$1/4$	1	$1/20$	0	20
a_1	1	0	$-3/25$	$4/25$	24
c^T	-25	0	4	0	1600

$$R_1 \leftrightarrow R_2$$

```
>>A=exchangeop(A,1,2)
```

	a_1	a_2	a_3	a_4	b
a_1	1	0	$-3/25$	$4/25$	24
a_2	$1/4$	1	$1/20$	0	20
c^T	-25	0	4	0	1600

$$R_2 \to R_2 - \tfrac{1}{4}R_1$$

```
>>A=eliminationop(A,2,1,-1/4)
```

	a_1	a_2	a_3	a_4	b
a_1	1	0	$-3/25$	$4/25$	24
a_2	0	1	$2/25$	$-1/25$	14
c^T	-25	0	4	0	1600

$$R_3 \to R_3 + 25R_1$$

```
>>A=eliminationop(A,3,1,25)
```

	a_1	a_2	a_3	a_4	b
a_1	1	0	$-3/25$	$4/25$	24
a_2	0	1	$2/25$	$-1/25$	14
c^T	0	0	1	4	2200

There is no negative element in the last row corresponding to non-basic variables. Therefore, we achieved optimal solution. That is,

$$B = [a_1 \quad a_2], x_B = \begin{bmatrix} x_1 \\ x_2 \end{bmatrix} = \begin{bmatrix} 24 \\ 14 \end{bmatrix}, x = \begin{bmatrix} 24 \\ 14 \\ 0 \\ 0 \end{bmatrix} \text{ and maximum value}$$

of an objective function is 2200.

Example 5.10. Solve the linear programming problem.

$$
\begin{aligned}
\text{maximize} \quad & 2x_1 + x_2 + 2x_3 + 9x_4 \\
\text{subject to} \quad & x_1 \qquad\qquad\quad +2x_4 = 2, \\
& x_2 - x_3 + x_4 = 4, \\
& x_1, \quad x_2, \quad x_3, \quad x_4 \geq 0.
\end{aligned}
$$

The standard form of the linear programming problem is given as

$$
\begin{aligned}
\text{minimize} \quad & -2x_1 - x_2 - 2x_3 - 9x_4 \\
\text{subject to} \quad & x_1 \qquad\qquad + 2x_4 = 2, \\
& x_2 - x_3 + x_4 = 4, \\
& x_1, \quad x_2, \quad x_3, \quad x_4 \geq 0.
\end{aligned}
$$

The augmented form of the matrix is

	a_1	a_2	a_3	a_4	b
	1	0	0	2	2
	0	1	−1	1	4
c^T	−2	−1	−2	−9	0

We have basis matrix $B = [a_1 \quad a_2]$, $x_B = \begin{bmatrix} x_1 \\ x_2 \end{bmatrix}$ and $Bx_B = b$ im-

plies $x_B = \begin{bmatrix} 2 \\ 4 \end{bmatrix}$. Therefore, the starting basic feasible solution is

$x = \begin{bmatrix} 2 \\ 4 \\ 0 \\ 0 \end{bmatrix}$. Note that there are four nonbasic variables, therefore we

take v=4.

```
>> v = 4
>> [A,q] = pivotcolumn(A,v)
```

	a_1	a_2	a_3	a_4	b
a_1	1	0	0	2	2
a_2	0	1	-1	1	4
c^T	-2	-1	-2	-9	0

q = 4.

We compute pivot row p = arg $\min\left\{\frac{2}{2}, \frac{4}{1}\right\}$ = 1st row.

```
>> B = [1  2]
>>[A,p,e,B]=pivotrow(A,q,B)
```

	a_1	a_2	a_3	a_4	b
a_1	1	0	0	2	2
a_2	0	1	-1	1	4
c^T	-2	-1	-2	-9	0

p = 1   e = 2   B = $\begin{bmatrix} 4 & 2 \end{bmatrix}$.

	a_1	a_2	a_3	a_4	b
a_4	1	0	0	2	2
a_2	0	1	-1	1	4
c^T	-2	-1	-2	-9	0

$$R_1 \to \tfrac{1}{2}R_1$$

```
>> A = identityop(A,p,1/2)
```

	a_1	a_2	a_3	a_4	b
a_4	$1/2$	0	0	1	1
a_2	0	1	-1	1	4
c^T	-2	-1	-2	-9	0

$$R_2 \to R_2 - R_1$$

```
>>A=eliminationop(A,2,p,-1)
```

$$
\begin{array}{cccccc}
 & a_1 & a_2 & a_3 & a_4 & b \\
a_4 & 1/2 & 0 & 0 & \boxed{1} & 1 \\
a_2 & -1/2 & 1 & -1 & 0 & 3 \\
c^T & -2 & -1 & -2 & -9 & 0
\end{array}
$$

$$R_3 \to R_3 + 9R_1$$

```
>> A = eliminationop(A,3,p,9)
```

$$
\begin{array}{cccccc}
 & a_1 & a_2 & a_3 & a_4 & b \\
a_4 & 1/2 & 0 & 0 & \boxed{1} & 1 \\
a_2 & -1/2 & 1 & -1 & 0 & 3 \\
c^T & 5/2 & -1 & -2 & 0 & 9
\end{array}
$$

$$R_1 \leftrightarrow R_2$$

```
>> A = exchangeop(A,1,2)
```

$$
\begin{array}{cccccc}
 & a_1 & a_2 & a_3 & a_4 & b \\
a_2 & -1/2 & 1 & \boxed{-1} & 0 & 3 \\
a_4 & 1/2 & 0 & 0 & 1 & 1 \\
c^T & 5/2 & -1 & -2 & 0 & 9
\end{array}
$$

```
>> [A,q] = pivotcolumn(A,v)
>> [A,p,e,B] = pivotrow(A,q,B)
```

$$
\begin{array}{cccccc}
 & a_1 & a_2 & a_3 & a_4 & b \\
a_2 & -1/2 & 1 & \boxed{-1} & 0 & 3 \\
a_4 & 1/2 & 0 & 0 & 1 & 1 \\
c^T & 5/2 & -1 & -2 & 0 & 9
\end{array}
$$

```
q = 3     unbounded.
```

We have $p = \arg\min\left\{\frac{\cancel{3}}{\cancel{-1}}, \frac{\cancel{1}}{\cancel{0}}\right\}$.=no row. There is no such p. Hence, the problem is unbounded.

Example 5.11. x_1 and x_2 are two positive real numbers such that $2x_1 + x_2 \le 6$ and $x_1 + 2x_2 \le 8$. For which of the following value of (x_1, x_2) the function $f(x_1, x_2) = 3x_1 + 6x_2$ will give the maximum value?

(a) $(4/3, 10/3)$

(b) $(8/3, 20/3)$

(c) $(8/3, 10/3)$

(d) $(4/3, 20/3)$

Given that

$$2x_1 + x_2 \le 6,$$
$$x_1 + 2x_2 \le 8.$$

We write the above inequalities into equation form

$$2x_1 + x_2 = 6 \text{ and } x_1 + 2x_2 = 8.$$

Solving the above two equations, we get

$$x_1 = \frac{4}{3} \text{ and } x_2 = \frac{10}{3}.$$

Therefore, $3x_1 + 6x_2$ will give maximum value at $(\frac{4}{3}, \frac{10}{3})$. Thus, option (a) is true.

Example 5.12. Which of the following statement is TRUE?

(a) A convex set cannot have infinite many extreme points.

(b) A linear programming problem can have infinite many extreme points.

(c) A linear programming problem can have exactly two different optimal solutions.

(d) A linear programming problem can have a nonbasic optimal solution.

A linear programming problem can have a nonbasic optimal solution. Thus, option (d) is true.

5.5   Two-Phase Simplex Method

Sometimes, we get a linear programming problem in which a starting basic feasible solution is not available and we cannot initiate the simplex algorithm. Therefore, we require a systematic method to find a starting basic feasible solution of such linear programming problems, so that the simplex method could be started. The two-phase method is useful in such situations.

Consider a linear programming problem

$$\text{minimize} \quad 2x_1 + 3x_2$$
$$\text{subject to} \quad 4x_1 + 2x_2 \geq 12,$$
$$x_1 + 4x_2 \geq 6,$$
$$x_1, \quad x_2 \geq 0.$$

We write the standard form as

$$\text{minimize} \quad 2x_1 + 3x_2$$
$$\text{subject to} \quad 4x_1 + 2x_2 - x_3 \qquad = 12,$$
$$x_1 + 4x_2 \qquad -x_4 = 6,$$
$$x_1, \quad x_2, \quad x_3, \quad x_4 \geq 0.$$

The tableau of the above problem is

	a_1	a_2	a_3	a_4	b
	4	2	-1	0	12
	1	4	0	-1	6
c^T	2	3	0	0	0

Since there is no basis matrix in the present tableau, we do not have any basic solution, and therefore, there is no basic feasible solution. Thus, we cannot initiate the simplex algorithm.

To proceed further, we consider an artificial problem as follows:

$$\text{minimize} \quad y_1 + y_2 + \cdots + y_n$$

$$\text{subject to} \quad \begin{bmatrix} A & I_m \end{bmatrix} \begin{bmatrix} x \\ y \end{bmatrix} = b,$$

$$\begin{bmatrix} x \\ y \end{bmatrix} \geq 0.$$

Given that $y = \begin{bmatrix} y_1 \\ y_2 \\ \vdots \\ y_n \end{bmatrix}$ is the vector of artificial variables. Note that the artificial problem has an obvious initial basic feasible solution $\begin{bmatrix} 0 \\ b \end{bmatrix}$.

Theorem 5.5. *The original linear programming problem has a basic feasible solution if and only if the associated artificial problem has an optimal feasible solution with objective function value zero.*

Proof. If the original problem has a basic feasible solution x, then the vector $\begin{bmatrix} x \\ 0 \end{bmatrix}$ is a basic feasible solution of the artifical problem. This basic feasible solution clearly gives an objective function value zero. Therefore, this solution is optimal for the artificial problem. Conversely, suppose that the artificial problem has an optimal feasible solution with objective function value zero. Then, this solution has the form $\begin{bmatrix} x \\ 0 \end{bmatrix}$, $x \geq 0$. It means that $Ax = b$ and x is feasible solution to original linear programming problem. By the fundamental theorem of linear programming problem (5.2), there also exists a basic feasible solution. $\square$

The two-phase method consists of phase I and phase II. In phase I, artificial variables are introduced in constraints and an objective function is formed using artificial variables only. The artificial variables have no meaning in a physical sense, but are useful to get the starting basic feasible solution of the linear programming problem.

We update the tableau to get the value of objective function as zero. If it is zero, then a basic feasible solution is available and we move to phase II; otherwise, it is determined that no feasible solutions exist and we stop in phase I.

In phase II, the artifical variables and the objective function of phase I are omitted and the original objective function is minimized using the basic feasible solution resulting from phase I and updating the tableau.

We illustrate the two-phase method in the following example.

Example 5.13. Solve the following linear programming problem using the two-phase method.

$$\begin{array}{ll} \text{minimize} & 2x_1 + 3x_2 \\ \text{subject to} & 4x_1 + 2x_2 \geq 12, \\ & x_1 + 4x_2 \geq 6, \\ & x_1, \quad x_2 \geq 0. \end{array}$$

The standard form of linear programming problem is given as

$$\begin{array}{ll} \text{minimize} & 2x_1 + 3x_2 \\ \text{subject to} & 4x_1 + 2x_2 - x_3 \quad\quad = 12, \\ & x_1 + 4x_2 \quad\quad -x_4 = 6, \\ & x_1, \quad x_2, \quad x_3, \quad x_4 \geq 0. \end{array}$$

This linear programming problem has no obvious feasible solution to use the simplex method. Therefore, we use the two-phase method.

Phase I: We introduce artificial variables. Consider an artificial problem:

$$\begin{array}{ll} \text{minimize} & x_5 + x_6 \\ \text{subject to} & 4x_1 + 2x_2 - x_3 \quad +x_5 \quad\quad = 12, \\ & x_1 + 4x_2 \quad\quad -x_4 \quad +x_6 = 6, \\ & x_1, \quad x_2, \quad x_3, \quad x_4, \quad x_5, \quad x_6 \geq 0. \end{array}$$

We form the tableau for the corresponding problem:

	a_1	a_2	a_3	a_4	a_5	a_6	b
a_5	4	2	-1	0	1	0	12
a_6	1	4	0	-1	0	1	6
c^T	0	0	0	0	1	1	0

Since basis matrix $B = \begin{bmatrix} a_5 & a_6 \end{bmatrix}$, therefore $x_B = \begin{bmatrix} x_5 \\ x_6 \end{bmatrix} = \begin{bmatrix} 12 \\ 6 \end{bmatrix}$.
We are looking for the basic feasible solutions which are $x_1 = 0$, $x_2 = 0$. To initiate the simplex method, we must update the last row of this tableau so that it has zero component under the basic variables.

We update the last row to get the feasible tableau. Updating the last row means that we subtract elements of the last row from the sum of the corresponding elements of rows where artificial variables are basis. That is,

$$R_3 \to R_3 - (R_1 + R_2)$$

	a_1	a_2	a_3	a_4	a_5	a_6	b
a_5	4	2	-1	0	1	0	12
a_6	1	4	0	-1	0	1	6
c^T	-5	-6	1	1	0	0	-18

The basic feasible solution of the above tableau is not optimal. Therefore, we proceed with the simplex method. We choose the most negative entry in the last row of the canonical tableau.

```
>> v = 4
>> [A,q] = pivotcolumn(A,v)
```

	a_1	a_2	a_3	a_4	a_5	a_6	b
a_5	4	2	-1	0	1	0	12
a_6	1	4	0	-1	0	1	6
c^T	-5	-6	1	1	0	0	-18

```
q = 2.
```

```
>> B = [5  6]
>>[A,p,e,B] = pivotrow(A,q,B)
```

$$
\begin{array}{c|cccccc|c}
 & a_1 & a_2 & a_3 & a_4 & a_5 & a_6 & b \\
a_5 & 4 & \boxed{2} & -1 & 0 & 1 & 0 & 12 \\
a_6 & \boxed{1} & 4 & 0 & -1 & 0 & 1 & 6 \\
c^T & -5 & \boxed{-6} & 1 & 1 & 0 & 0 & -18
\end{array}
$$

p = 2   e = 4   B = $\begin{bmatrix} 5 & 2 \end{bmatrix}$.

$$
\begin{array}{c|ccccccc}
 & a_1 & a_2 & a_3 & a_4 & a_5 & a_6 & b \\
a_5 & 4 & 2 & -1 & 0 & 1 & 0 & 12 \\
a_2 & 1 & \boxed{4} & 0 & -1 & 0 & 1 & 6 \\
c^T & -5 & -6 & 1 & 1 & 0 & 0 & -18
\end{array}
$$

$R_2 \rightarrow \frac{1}{4}R_2$

$$
\begin{array}{c|ccccccc}
 & a_1 & a_2 & a_3 & a_4 & a_5 & a_6 & b \\
a_5 & 4 & 2 & -1 & 0 & 1 & 0 & 12 \\
a_2 & 1/4 & \boxed{1} & 0 & -1/4 & 0 & 1/4 & 3/2 \\
c^T & -5 & -6 & 1 & 1 & 0 & 0 & -18
\end{array}
$$

$R_1 \rightarrow R_1 - 2R_2$

$$
\begin{array}{c|ccccccc}
 & a_1 & a_2 & a_3 & a_4 & a_5 & a_6 & b \\
a_5 & 7/2 & 0 & -1 & 1/2 & 1 & -1/2 & 9 \\
a_2 & 1/4 & \boxed{1} & 0 & -1/4 & 0 & 1/4 & 3/2 \\
c^T & -5 & -6 & 1 & 1 & 0 & 0 & -18
\end{array}
$$

$R_3 \rightarrow R_3 + 6R_2$

	a_1	a_2	a_3	a_4	a_5	a_6	b
a_5	$7/2$	0	-1	$1/2$	1	$-1/2$	9
a_2	$1/4$	1	0	$-1/4$	0	$1/4$	$3/2$
c^T	$-7/2$	0	1	$-1/2$	0	$3/2$	-9

The most negative element for the pivot column is q=1 and the pivot row is p=arg min$\left\{ \frac{9}{7/2}, \frac{3/2}{1/4} \right\} = 1^{\text{st}}$ row.

```
>> [A,q] = pivotcolumn(A,v)
>> B = [5  2]
>> [A,p,e,B] = pivotrow(A,q,B)
```

	a_1	a_2	a_3	a_4	a_5	a_6	b
a_5	$7/2$	0	-1	$1/2$	1	$-1/2$	9
a_2	$1/4$	1	0	$-1/4$	0	$1/4$	$3/2$
c^T	$-7/2$	0	1	$-1/2$	0	$3/2$	-9

p = 1   q = 1   e = $7/2$ B = $\begin{bmatrix} 1 & 2 \end{bmatrix}$.

	a_1	a_2	a_3	a_4	a_5	a_6	b
a_1	$7/2$	0	-1	$1/2$	1	$-1/2$	9
a_2	$1/4$	1	0	$-1/4$	0	$1/4$	$3/2$
c^T	$-7/2$	0	1	$-1/2$	0	$3/2$	-9

$R_1 \to \frac{2}{7} R_1$

	a_1	a_2	a_3	a_4	a_5	a_6	b
a_1	1	0	$-2/7$	$1/7$	$2/7$	$-1/7$	$18/7$
a_2	$1/4$	1	0	$-1/4$	0	$1/4$	$3/2$
c^T	$-7/2$	0	1	$-1/2$	0	$3/2$	-9

$R_2 \to R_2 - \frac{1}{4} R_1$

	a_1	a_2	a_3	a_4	a_5	a_6	b
a_1	1	0	$-2/7$	$1/7$	$2/7$	$-1/7$	$18/7$
a_2	0	1	$1/14$	$-2/7$	$-1/14$	$2/7$	$6/7$
c^T	$-7/2$	0	1	$-1/2$	0	$3/2$	-9

$R_3 \rightarrow R_3 + \frac{7}{2}R_1$

	a_1	a_2	a_3	a_4	a_5	a_6	b
a_1	1	0	$-2/7$	$1/7$	$2/7$	$-1/7$	$18/7$
a_2	0	1	$1/14$	$-2/7$	$-1/14$	$2/7$	$6/7$
c^T	0	0	0	0	1	1	0

Note that both the artificial variables have been driven out of the basis and the current basic feasible solution $x = \begin{bmatrix} 18/7 \\ 6/7 \end{bmatrix}$ is giving objective value zero to the artificial problem.

Phase II: We apply the simplex algorithm to the original linear programming problem after deleting the columns corresponding to the artificial variables and writing the cost of the original problem.

	a_1	a_2	a_3	a_4	b
a_1	1	0	$-2/7$	$1/7$	$18/7$
a_2	0	1	$1/14$	$-2/7$	$6/7$
c^T	2	3	0	0	0

$R_3 \rightarrow R_3 - 2R_1$

	a_1	a_2	a_3	a_4	b
a_1	1	0	$-2/7$	$1/7$	$18/7$
a_2	0	1	$1/14$	$-2/7$	$6/7$
c^T	0	3	$4/7$	$-2/7$	$-36/7$

$R_3 \rightarrow R_3 - 3R_2$

	a_1	a_2	a_3	a_4	b
a_1	1	0	$-2/7$	$1/7$	$18/7$
a_2	0	1	$1/14$	$-2/7$	$6/7$
c^T	0	0	$5/14$	$4/7$	$-54/7$

The cost corresponding to basic variables should be zero. We get basic variables x_1 and x_2. Since all costs are non-negative, therefore the current basic feasible solution is optimal. That is, $x_1 = \frac{18}{7}$, $x_2 = \frac{6}{7}$ and $x = \begin{bmatrix} 18/7 \\ 6/7 \\ 0 \\ 0 \end{bmatrix}$. The minimum value of an objective function is $\frac{54}{7}$.

We can take the advantage of MATLAB code `updatelastrow.m` given in the following **Code 5.4** to get the feasible tableau.

Code 5.4: updatelastrow.m

```
function A=updatelastrow(A,av)
%input : augmented matrix A, artifical
%          variables av
%output: augmented matrix
%update last row of tableau
[m,n]=size(A);
sum=0;
for J=1:n
    for I=1:av
        sum=sum+A(I,J);
    end
    A(m,J)=A(m,J)-sum;
    sum=0;
end
return
```

Example 5.14. Solve the linear programming problem using the two-phase method.

$$\begin{aligned}
\text{minimize} \quad & 4x_1 + x_2 + x_3 \\
\text{subject to} \quad & 2x_1 + x_2 + 2x_3 = 4, \\
& 3x_1 + 3x_2 + x_3 = 3, \\
& x_1, \quad x_2, \quad x_3 \geq 0.
\end{aligned}$$

a_1	a_2	a_3	b
2	1	2	4
3	3	1	3
4	1	1	0

There is no basic feasible solution. Therefore, we move to the two-phase simplex method.

Phase I:

We introduce artificial variables $x_4 \geq 0$, $x_5 \geq 0$ and an objective function $x_4 + x_5$. Our linear programming problem will be in the form:

$$\begin{aligned}
\text{minimize} \quad & x_4 + x_5 \\
\text{subject to} \quad & 2x_1 + x_2 + 2x_3 + x_4 = 4, \\
& 3x_1 + 3x_2 + x_3 + x_5 = 3, \\
& x_1, \quad x_2, \quad x_3, \quad x_4, \quad x_5 \geq 0.
\end{aligned}$$

We use MATLAB function **updatelastrow.m** to update the last row of tableau. Note that the number of artificial variables is 2. Therefore, we take `av=2`.

	a_1	a_2	a_3	a_4	a_5	b
a_4	2	1	2	1	0	4
a_5	3	3	1	0	1	3
c^T	0	0	0	1	1	0

$$R_3 \to R_3 - (R_1 + R_2)$$

```
>> av = 2
>> A = updatelastrow(A,av)
```

	a_1	a_2	a_3	a_4	a_5	b
a_4	2	1	2	1	0	4
a_5	3	3	1	0	1	3
c^T	-5	-4	-3	0	0	-7

We get the feasible tableau. Thus, we can apply the simplex method.

```
>> v = 3
>> [A,q] = pivotcolumn(A,v)
```

	a_1	a_2	a_3	a_4	a_5	b
a_4	2	1	2	1	0	4
a_5	3	3	1	0	1	3
c^T	−5	−4	−3	0	0	−7

```
  q = 1.
```

We find pivot row $p = \arg\min\left\{\frac{4}{2}, \frac{3}{3}\right\} = 2^{\text{nd}}$ row.

```
>> B = [4 5]
>> [A,p,e,B] = pivotrow(A,q,B)
```

	a_1	a_2	a_3	a_4	a_5	b
a_4	2	1	2	1	0	4
a_5	3	3	1	0	1	3
c^T	−5	−4	−3	0	0	−7

```
  p = 2   e = 3   B = [4 1].
```

	a_1	a_2	a_3	a_4	a_5	b
a_4	2	1	2	1	0	4
a_1	3	3	1	0	1	3
c^T	−5	−4	−3	0	0	−7

$$R_2 \to \tfrac{1}{3} R_2$$

```
>> A = identityop(A,p,1/3)
```

	a_1	a_2	a_3	a_4	a_5	b
a_4	2	1	2	1	0	4
a_1	1	1	$1/3$	0	$1/3$	1
c^T	−5	−4	−3	0	0	−7

$$R_1 \to R_1 - 2R_2$$

```
>> A = eliminationop(A,1,p,-2)
```

	a_1	a_2	a_3	a_4	a_5	b
a_4	0	-1	$4/3$	1	$-2/3$	2
a_1	$\boxed{1}$	1	$1/3$	0	$1/3$	1
c^T	-5	-4	-3	0	0	-7

$$R_3 \to R_3 + 5R_2$$

```
>> A = eliminationop(A,3,p,5)
```

	a_1	a_2	a_3	a_4	a_5	b
a_4	0	-1	$4/3$	1	$-2/3$	2
a_1	1	1	$1/3$	0	$1/3$	1
c^T	0	1	$-4/3$	0	$5/3$	-2

```
>> [A,q] = pivotcolumn(A,v)
```

	a_1	a_2	a_3	a_4	a_5	b
a_4	0	-1	$4/3$	1	$-2/3$	2
a_1	1	1	$1/3$	0	$1/3$	1
c^T	0	1	$-4/3$	0	$5/3$	-2

q = 3.

To update the simplex tableau, we calculate pivot row p = arg min $\left\{ \frac{2 \times 3}{4}, \frac{1 \times 3}{1} \right\}$ = arg min $\left\{ \frac{3}{2}, 3 \right\}$ = 1st row.

```
>> [A,p,e,B] = pivotrow(A,q,B)
```

	a_1	a_2	a_3	a_4	a_5	b
a_4	0	-1	$4/3$	1	$-2/3$	2
a_1	1	1	$1/3$	0	$1/3$	1
c^T	0	1	$-4/3$	0	$5/3$	-2

p=1   e = $4/3$ B = $\begin{bmatrix} 3 & 1 \end{bmatrix}$.

	a_1	a_2	a_3	a_4	a_5	b
a_3	0	-1	$\boxed{4/3}$	1	$-2/3$	2
a_1	1	1	$1/3$	0	$1/3$	1
c^T	0	1	$-4/3$	0	$5/3$	-2

$$R_1 \to \tfrac{3}{4}R_1$$

```
>> A = identityop(A,p,3/4)
```

	a_1	a_2	a_3	a_4	a_5	b
a_3	0	$-3/4$	$\boxed{1}$	$3/4$	$-1/2$	$3/2$
a_1	1	1	$1/3$	0	$1/3$	1
c^T	0	1	$-4/3$	0	$5/3$	-2

$$R_2 \to R_2 - \tfrac{1}{3}R_1$$

```
>>A=eliminationop(A,2,p,-1/3)
```

	a_1	a_2	a_3	a_4	a_5	b
a_3	0	$-3/4$	$\boxed{1}$	$3/4$	$-1/2$	$3/2$
a_1	1	$5/4$	0	$-1/4$	$1/2$	$1/2$
c^T	0	1	$-4/3$	0	$5/3$	-2

$$R_3 \to R_3 + \tfrac{4}{3}R_1$$

```
>>A=eliminationop(A,3,p,4/3)
```

	a_1	a_2	a_3	a_4	a_5	b
a_3	0	$-3/4$	$\boxed{1}$	$3/4$	$-1/2$	$3/2$
a_1	1	$5/4$	0	$-1/4$	$1/2$	$1/2$
c^T	0	0	0	1	1	0

$$R_1 \leftrightarrow R_2$$

```
>> A = exchangeop(A,1,2)
```

	a_1	a_2	a_3	a_4	a_5	b
a_1	1	$5/4$	0	$-1/4$	$1/2$	$1/2$
a_3	0	$-3/4$	1	$3/4$	$-1/2$	$3/2$
c^T	0	0	0	1	1	0

Since the basis matrix is $B = \begin{bmatrix} a_1 & a_3 \end{bmatrix}$ and $x_B = \begin{bmatrix} x_1 \\ x_3 \end{bmatrix}$, therefore,

$x = \begin{bmatrix} 1/2 \\ 0 \\ 3/2 \\ 0 \\ 0 \end{bmatrix}$. The value of the artificial problem is zero. Therefore,

we proceed to phase II.

Phase II: We start by deleting the columns corresponding to artificial variables and using costs of the original problem.

	a_1	a_2	a_3	b
a_1	1	$5/4$	0	$1/2$
a_3	0	$-3/4$	1	$3/2$
c^T	4	1	1	0

$$R_3 \to R_3 - 4R_1$$

```
>> A = eliminationop(A,3,p,-4)
```

	a_1	a_2	a_3	b
a_1	1	$5/4$	0	$1/2$
a_3	0	$-3/4$	1	$3/2$
c^T	0	-4	1	-2

$$R_3 \to R_3 - R_2$$

```
>> A = eliminationop(A,3,2,-1)
```

	a_1	a_2	a_3	b
a_1	1	$5/4$	0	$1/2$
a_3	0	$-3/4$	1	$3/2$
c^T	0	$-13/4$	0	$-7/2$

```
>> [A,q] = pivotcolumn(A,v)
```

	a_1	a_2	a_3	b
a_1	1	$5/4$	0	$1/2$
a_3	0	$-3/4$	1	$3/2$
c^T	0	$-13/4$	0	$-7/2$

```
  q = 2.
>> B= [1  3]
>> [A,p,e,B] = pivotrow(A,q,B)
```

	a_1	a_2	a_3	b
a_1	1	$5/4$	0	$1/2$
a_3	0	$-3/4$	1	$3/2$
c^T	0	$-13/4$	0	$-7/2$

```
  p = 1   e = 5/4   B = [2  3].
```

	a_1	a_2	a_3	b
a_2	1	$5/4$	0	$1/2$
a_3	0	$-3/4$	1	$3/2$
c^T	0	$-13/4$	0	$-7/2$

$$R_1 \to \tfrac{4}{5}R_1$$

```
>> A = identityop(A,p,4/5)
```

$$
\begin{array}{ccccc}
 & a_1 & a_2 & a_3 & b \\
a_2 & 4/5 & \boxed{1} & 0 & 2/5 \\
a_3 & 0 & -3/4 & 1 & 3/2 \\
c^T & 0 & -13/4 & 0 & -7/2 \\
\end{array}
$$

$$R_2 \to R_2 + \tfrac{3}{4}R_1$$

```
>> A = eliminationop(A,2,p,3/4)
```

$$
\begin{array}{ccccc}
 & a_1 & a_2 & a_3 & b \\
a_2 & 4/5 & \boxed{1} & 0 & 2/5 \\
a_3 & 3/5 & 0 & 1 & 9/5 \\
c^T & 0 & -13/4 & 0 & -7/2 \\
\end{array}
$$

$$R_3 \to R_3 + \tfrac{13}{4}R_1$$

```
>> A = eliminationop(A,3,p,13/4)
```

$$
\begin{array}{ccccc}
 & a_1 & a_2 & a_3 & b \\
a_2 & 4/5 & 1 & 0 & 2/5 \\
a_3 & 3/5 & 0 & 1 & 9/5 \\
c^T & 13/5 & 0 & 0 & -11/5 \\
\end{array}
$$

All costs are non-negative. Since basis matrix $B = \begin{bmatrix} a_2 & a_3 \end{bmatrix}$ and basic vector $x_B = \begin{bmatrix} x_2 \\ x_3 \end{bmatrix}$, therefore $x = \begin{bmatrix} 0 \\ 2/5 \\ 9/5 \end{bmatrix}$ and an optimal value is $\frac{11}{5}$.

5.6 Exercises

Exercise 5.1. Find all basic feasible solutions of the following system of linear equations.

(a)
$$x_1 - x_2 + 2x_3 = 18,$$
$$x_1 + 2x_2 - x_3 = 1.$$

(b)
$$x_1 + x_2 + x_3 = 1,$$
$$3x_1 + 2x_2 - x_4 = 6.$$

Exercise 5.2. Find all basic feasible solutions of the following system of linear equations.

$$x_1 + 2x_2 + 4x_3 + x_4 = 7,$$
$$2x_1 - x_2 + 3x_3 - 2x_4 = 4.$$

Indicate which of these solutions are feasible.

Exercise 5.3. Solve the following linear programming problem by the simplex method.

$$
\begin{aligned}
\text{maximize} \quad & 3x_1 + 4x_2 + 4x_3 + 7x_4 \\
\text{subject to} \quad & 8x_1 + 3x_2 + 4x_3 + x_4 \leq 7, \\
& 2x_1 + 6x_2 + x_3 + 5x_4 \leq 10, \\
& x_1 + 4x_2 + 5x_3 + 2x_4 \leq 8, \\
& x, \quad x_2, \quad x_3, \quad x_4 \geq 0.
\end{aligned}
$$

Exercise 5.4. Solve the following problem by the simplex method.

$$
\begin{aligned}
\text{maximize} \quad & 11x_1 + 10x_2 \\
\text{subject to} \quad & -0.5x_1 + 1.3x_2 \leq 0.8, \\
& 4x_1 + x_2 \leq 12.7, \\
& 6x_1 + x_2 \leq 15.4, \\
& 6x_1 - x_2 \leq 13.4, \\
& 4x_1 - x_2 \leq 8.7, \\
& 5x_1 - 3x_2 \leq 10.0. \\
& x_1, \quad x_2 \geq 0.
\end{aligned}
$$

Exercise 5.5. Solve the following problem by the simplex method.

$$
\begin{aligned}
x_1 + x_2 &\leq 3, \\
x_1 - 2x_2 &\leq 1, \\
-2x_1 + x_2 &\leq 2, \\
x_1, \quad x_2 &\geq 0.
\end{aligned}
$$

(a) maximize $x_1 - x_2$

(b) minimize $x_1 - x_2$

Exercise 5.6. Solve the problem by the simplex method.

$$
\begin{aligned}
\text{maximize} \quad & 7x_1 + 5x_2 \\
\text{subject to} \quad & x_1 + 2x_2 \leq 6, \\
& 4x_1 + 3x_2 \leq 12, \\
& x_1, \quad x_2 \geq 0.
\end{aligned}
$$

Exercise 5.7. Solve the problem by the simplex method.

$$
\begin{aligned}
\text{maximize} \quad & 9x_1 + 7x_2 \\
\text{subject to} \quad & x_1 + 2x_2 \leq 7, \\
& x_1 - x_2 \leq 4, \\
& x_1, \quad x_2 \geq 0.
\end{aligned}
$$

Exercise 5.8. Solve the linear programming problem using the simplex method.

$$
\begin{aligned}
\text{maximize} \quad & 10x_1 + 6x_2 - 8x_3 \\
\text{subject to} \quad & 5x_1 - 2x_2 + 6x_3 \leq 20, \\
& 10x_1 + 4x_2 - 6x_3 \leq 30, \\
& x_1, \quad x_2, \quad x_3 \geq 0.
\end{aligned}
$$

Exercise 5.9. Solve the problem by the simplex method.

$$
\begin{aligned}
\text{maximize} \quad & 2x_1 + x_2 \\
\text{subject to} \quad & 2x_1 + 3x_2 \leq 3, \\
& x_1 + 5x_2 \leq 2, \\
& 2x_1 + x_2 \leq 5, \\
& x_1, \quad x_2 \geq 0.
\end{aligned}
$$

Exercise 5.10. Solve the linear programming problem using the simplex method.

$$\begin{aligned}
\text{maximize} \quad & 5x_1 + x_2 \\
\text{subject to} \quad & 3x_1 - 2x_2 \le 6, \\
& -4x_1 + 2x_2 \le 4, \\
& x_1, \quad x_2 \ge 0.
\end{aligned}$$

Exercise 5.11. Solve the linear programming problem using the two-phase method.

$$\begin{aligned}
\text{minimize} \quad & 2x_1 + 4x_2 + 7x_3 + x_4 + 5x_5 \\
\text{subject to} \quad -\; & x_1 + x_2 + 2x_3 + x_4 + 2x_5 = 7, \\
-\; & x_1 + 2x_2 + 3x_3 + x_4 + x_5 = 6, \\
-\; & x_1 + x_2 + x_3 + 2x_4 + x_5 = 4, \\
& x_1, \quad x_2, \quad x_3, \quad x_4, \quad x_5 \ge 0.
\end{aligned}$$

Exercise 5.12. Solve the following linear programming problem using the two-phase method.

$$\begin{aligned}
\text{minimize} \quad & 5x_1 + 8x_2 \\
\text{subject to} \quad & 3x_1 + 2x_2 \ge 3, \\
& x_1 + 4x_2 \ge 4, \\
& x_1 + x_2 \ge 5, \\
& x_1, \quad x_2 \ge 0.
\end{aligned}$$

Chapter 6

The Revised Simplex Method

6.1 Introduction

Consider a linear programming problem in standard form with a simplex tableau A of size $m \times n$. Suppose that we wish to solve this problem using the simplex method. Experience suggests that if the simplex tableau A has fewer rows m than columns n, then in most instances, pivots occur in only a small fraction of the columns of the simplex tableau A. The operation of pivoting involves updating all the columns of the simplex tableau to move from one iteration to next in search of an improved solution. However, if a particular column of A never enters into basis during the entire simplex procedure, then computations performed on this column are not explicitly used. Therefore, the effort expended on performing operations on many such columns of A may be a waste. The revised simplex method reduces the amount of computation leading to an optimal solution by eliminating operations on columns of A that do not enter into the basis.

Therefore, we apply the revised simplex method to avoid the unnecessary calculations and save the computational time.

6.2 Matrix Form of the Revised Simplex Method

Consider a linear programming problem in standard form

$$\begin{aligned} \text{minimize} \quad & c^T x \\ \text{subject to} \quad & Ax = b, \\ & x \geq 0. \end{aligned}$$

Let the first m columns of A be the basic columns. The columns form a square $m \times m$ nonsingular matrix B. The nonbasic columns of A form an $m \times (n-m)$ matrix D. We correspondingly partition the cost vector as $c^T = [c_B^T, c_D^T]$. Then, the original linear programming problem can be represented as follows:

$$\begin{aligned} \text{minimize} \quad & c_B^T x_B + c_D^T x_D \\ \text{subject to} \quad & Bx_B + Dx_D = b, \\ & x_B \qquad\qquad \geq 0, \\ & \qquad\qquad x_D \geq 0. \end{aligned}$$

If $x_D = 0$, then the solution $x = \begin{bmatrix} x_B^T, & x_D^T \end{bmatrix}^T = \begin{bmatrix} x_B^T, & 0^T \end{bmatrix}^T$ is the basic feasible solution corresponding to the basis matrix B. It is clear that for this to be the solution, we need $x_B = B^{-1}b$, that is, the basic feasible solution is

$$x = \begin{bmatrix} B^{-1}b \\ 0 \end{bmatrix}.$$

The corresponding objective function value is $z_0 = c_B^T B^{-1}b$.
But, if $x_D \neq 0$, then the solution $x = \begin{bmatrix} x_B^T, & x_D^T \end{bmatrix}$ is not basic. In this case, x_B is given by

$$x_B = B^{-1}b - B^{-1}Dx_D,$$

and the corresponding objective function value is

$$\begin{aligned} z &= c_B^T x_B + c_D^T x_D \\ &= c_B^T(B^{-1}b - B^{-1}Dx_D) + c_D^T c_D \\ &= c_B^T B^{-1}b + (c_D^T - c_B^T B^{-1}D)x_D. \end{aligned}$$

We can define $r_D^T = c_D^T - \lambda^T D$, where $\lambda^T = c_B^T B^{-1}$.
The elements of vector r_D^T are called the reduced cost coefficients corresponding to the nonbasic variables.
If $r_D^T \geq 0$, then the basic feasible solution corresponding to the basis B is optimal. If, on the other hand, any component of r_D^T is negative, then the value of the objective function can be reduced by increasing a corresponding component of x_D, that is, by changing the basis.
We now use the above observations in the revised simplex algorithm.

6.3 The Revised Simplex Algorithm

We have the following steps for the revised simplex algorithm.

Algorithm

1. Form a revised simplex tableau corresponding to a starting basic feasible solution $\begin{bmatrix} B^{-1} & b_0 \end{bmatrix}$.

2. Calculate the current reduced cost coefficient vector: $r_D^T = c_D^T - \lambda^T D$, where $\lambda^T = c_B^T B^{-1}$.

3. If $r_j \geq 0$ for all j, then the current basic feasible solution is optimal so stop, otherwise go to step 4.

4. Select a q such that $r_q < 0$ (i.e., the q corresponding to the most negative r_q).

5. Compute $y_q = B^{-1} a_q$.

6. If no $y_{iq} \geq 0$, then the problem is unbounded so stop, otherwise go to step 7.

7. Compute p $= \arg \min_i \left\{ \dfrac{b_{i0}}{y_{iq}} : y_{iq} > 0 \right\}$.

8. Form the augmented revised table $\begin{bmatrix} B^{-1} & b_0 & y_q \end{bmatrix}$ and update the augmented tableau pivoting about the pth element of the last column that is, y_q of the tableau.

9. Remove the last column that is, y_q of the augmented revised tableau and go to step 2.

We write the linear programming problem in an augmented matrix form. We form the revised tableau using B^{-1} and b_0. The reduced cost coefficients are calculated to find pivot column corresponding to the most negative element in step 2. If reduced cost coefficients are non-negative, then the problem is optimal, otherwise we form the augmented revised tableau in step 8. We compute p $= \arg \min \left\{ \dfrac{b_{i0}}{y_{iq}} : y_{iq} > 0 \right\}$ for pivot row in step 7. If all y_{iq} are negative in step 6, then the problem is unbounded. In every iteration, we obtain an updated revised tableau consisting of B^{-1} and

b_0 with removal of the last column that is, y_{iq} of the augmented revised tableau in step 9.

Example 6.1. Solve the linear programming problem using the revised simplex method.

$$
\begin{aligned}
\text{maximize} \quad & 3x_1 + x_2 + 3x_3 \\
\text{subject to} \quad & 2x_1 + x_2 + x_3 \le 2, \\
& x_1 + 2x_2 + 3x_3 \le 5, \\
& 2x_1 + 2x_2 + x_3 \le 6, \\
& x_1, \quad x_2, \quad x_3 \ge 0.
\end{aligned}
$$

We express the problem in standard form:

$$
\begin{aligned}
\text{minimize} \quad & -3x_1 - x_2 - 3x_3 \\
\text{subject to} \quad & 2x_1 + x_2 + x_3 + x_4 = 2, \\
& x_1 + 2x_2 + 3x_3 + x_5 = 5, \\
& 2x_1 + 2x_2 + x_3 + x_6 = 6, \\
& x_1, \quad x_2, \quad x_3, \quad x_4, \, x_5, \, x_6 \ge 0.
\end{aligned}
$$

The augmented matrix is

	a_1	a_2	a_3	a_4	a_5	a_6	b
a_4	2	1	1	1	0	0	2
a_5	1	2	3	0	1	0	5
a_6	2	2	1	0	0	1	6

Since the basis matrix is $B = \begin{bmatrix} a_4 & a_5 & a_6 \end{bmatrix}$, therefore basic vector is $x_B = \begin{bmatrix} x_4 \\ x_5 \\ x_6 \end{bmatrix} = \begin{bmatrix} 2 \\ 5 \\ 6 \end{bmatrix}$. We have basic feasible solution, that is

$x = \begin{bmatrix} 0 \\ 0 \\ 0 \\ 2 \\ 5 \\ 6 \end{bmatrix}$. Therefore, we can start the revised simplex method.

We determine

$$c_B^T = \begin{bmatrix} c_4 & c_5 & c_6 \end{bmatrix}, \ c_D^T = \begin{bmatrix} c_1 & c_2 & c_3 \end{bmatrix}$$

and

$$\lambda^T = c_B^T B^{-1} = \begin{bmatrix} 0 & 0 & 0 \end{bmatrix} \begin{bmatrix} 1 & 0 & 0 \\ 0 & 1 & 0 \\ 0 & 0 & 1 \end{bmatrix} = \begin{bmatrix} 0 & 0 & 0 \end{bmatrix}.$$

The revised simplex table is

$$\begin{array}{cccc} & B^{-1} & & b \\ a_4 & 1 & 0 & 0 & 2 \\ a_5 & 0 & 1 & 0 & 5 \\ a_6 & 0 & 0 & 1 & 6 \end{array}$$

We compute

$$r_D^T = c_D^T - \lambda^T D$$

$$= \begin{bmatrix} c_1 & c_2 & c_3 \end{bmatrix} - \begin{bmatrix} 0 & 0 & 0 \end{bmatrix} \begin{bmatrix} 2 & 1 & 1 \\ 1 & 2 & 3 \\ 2 & 7 & 1 \end{bmatrix}$$

$$= \begin{bmatrix} -3 & -1 & -3 \end{bmatrix}$$

$$= \begin{bmatrix} r_1 & r_2 & r_3 \end{bmatrix}.$$

We have to choose the most negative reduced cost coefficients; they are $r_1 = r_3 = -3$. We take the first one, i.e., a_1 and bring a_1 into the basis. However, we first calculate

$$y_1 = \begin{bmatrix} 1 & 0 & 0 \\ 0 & 1 & 0 \\ 0 & 0 & 1 \end{bmatrix} \begin{bmatrix} 2 \\ 1 \\ 2 \end{bmatrix} = \begin{bmatrix} 2 \\ 1 \\ 2 \end{bmatrix}.$$

We now form the augmented revised tableau

$$\begin{array}{ccccc} & B^{-1} & & b & y_1 \\ a_4 & 1 & 0 & 0 & 2 & 2 \\ a_5 & 0 & 1 & 0 & 5 & 1 \\ a_6 & 0 & 0 & 1 & 6 & 2 \end{array}$$

Then, we compute p $= \arg\min \left\{ \frac{2}{2}, \frac{5}{1}, \frac{6}{2} \right\} = 1^{st}$ row and apply elementary row operations by pivoting about the 1st element of the last column of the above revised tableau.

$$
\begin{array}{c c c c c c}
 & B^{-1} & & & b & y_1 \\
a_1 & 1 & 0 & 0 & 2 & \boxed{2} \\
a_5 & 0 & 1 & 0 & 5 & 1 \\
a_6 & 0 & 0 & 1 & 6 & 2
\end{array}
$$

$R_1 \to \frac{R_1}{2}$

$$
\begin{array}{c c c c c c}
 & B^{-1} & & & b & y_1 \\
a_1 & \frac{1}{2} & 0 & 0 & 1 & \boxed{1} \\
a_5 & 0 & 1 & 0 & 5 & 1 \\
a_6 & 0 & 0 & 1 & 6 & 2
\end{array}
$$

$R_2 \to R_2 - \ R_1$
$R_3 \to R_3 - 2R_1$

$$
\begin{array}{c c c c c c}
 & B^{-1} & & & b & y_1 \\
a_1 & \frac{1}{2} & 0 & 0 & 1 & 1 \\
a_5 & -\frac{1}{2} & 1 & 0 & 4 & 0 \\
a_6 & -1 & 0 & 1 & 4 & 0
\end{array}
$$

After removing the last column, that is y_1, we get

$$
\begin{array}{c c c c c}
 & B^{-1} & & & b \\
a_1 & \frac{1}{2} & 0 & 0 & 1 \\
a_5 & -\frac{1}{2} & 1 & 0 & 4 \\
a_6 & -1 & 0 & 1 & 4
\end{array}
$$

We compute

$$
\begin{aligned}
\lambda^T &= c_B^T B^{-1} \\
&= \begin{bmatrix} c_1 & c_5 & c_6 \end{bmatrix} B^{-1} \\
&= \begin{bmatrix} -3 & 0 & 0 \end{bmatrix} \begin{bmatrix} \frac{1}{2} & 0 & 0 \\ -\frac{1}{2} & 1 & 0 \\ -1 & 0 & 1 \end{bmatrix} \\
&= \begin{bmatrix} -\frac{3}{2} & 0 & 0 \end{bmatrix}
\end{aligned}
$$

and

$$r_D^T = c_D^T - \lambda^T D$$

$$= \begin{bmatrix} c_2 & c_3 & c_4 \end{bmatrix} - \begin{bmatrix} -3/2 & 0 & 0 \end{bmatrix} \begin{bmatrix} 1 & 1 & 1 \\ 2 & 3 & 0 \\ 2 & 1 & 0 \end{bmatrix}$$

$$= \begin{bmatrix} -1 & -3 & 0 \end{bmatrix} - \begin{bmatrix} -3/2 & -3/2 & -3/2 \end{bmatrix}$$

$$= \begin{bmatrix} 1/2 & -3/2 & 3/2 \end{bmatrix}$$

$$= \begin{bmatrix} r_2 & r_3 & r_4 \end{bmatrix}.$$

We observe that reduced cost coefficient r_3 is only negative. Therefore, we bring a_3 into the basis.

$$y_3 = B^{-1} a_3$$

$$= \begin{bmatrix} 1/2 & 0 & 0 \\ -1/2 & 1 & 0 \\ -1 & 0 & 1 \end{bmatrix} \begin{bmatrix} 1 \\ 3 \\ 1 \end{bmatrix} = \begin{bmatrix} 1/2 \\ -1/2 + 3 \\ -1 + 0 + 1 \end{bmatrix} = \begin{bmatrix} 1/2 \\ 5/2 \\ 0 \end{bmatrix}.$$

In this case, we get p= arg min $\left\{ 1 \times 2, \frac{4 \times 2}{5}, \cancel{\frac{4}{0}} \right\} = 2^{\text{nd}}$ row.

	B^{-1}			b	y_3
a_1	1/2	0	0	1	1/2
a_3	-1/2	1	0	4	5/2
a_6	-1	0	1	4	0

We update the revised simplex tableau by pivoting the 2nd element of the last column.

$R_2 \to \frac{2}{5} R_2$

	B^{-1}			b	y_3
a_1	1/2	0	0	1	1/2
a_3	-1/5	2/5	0	8/5	1
a_6	-1	0	1	4	0

$R_1 \to R_1 - \frac{1}{2} R_2$

	B^{-1}			b	y_3
a_1	3/5	-1/5	0	1/5	0
a_3	-1/5	2/5	0	8/5	1
a_6	-1	0	1	4	0

After removing the last column in the above tableau,

$$
\begin{array}{ccccc}
 & \multicolumn{3}{c}{B^{-1}} & b \\
a_1 & 3/5 & -1/5 & 0 & 1/5 \\
a_3 & -1/5 & 2/5 & 0 & 8/5 \\
a_6 & -1 & 0 & 1 & 4
\end{array}
$$

We again compute

$$
\begin{aligned}
\lambda^T &= c_B^T B^{-1} \\
&= \begin{bmatrix} c_1 & c_3 & c_6 \end{bmatrix} B^{-1} \\
&= \begin{bmatrix} -3 & -3 & 0 \end{bmatrix}
\begin{bmatrix}
3/5 & -1/5 & 0 \\
-1/5 & 2/5 & 0 \\
-1 & 0 & 1
\end{bmatrix} \\
&= \begin{bmatrix} \frac{-9}{5} + \frac{3}{5} & \frac{3}{5} - \frac{6}{5} & 0 \end{bmatrix} \\
&= \begin{bmatrix} \frac{-6}{5}, & \frac{-3}{5}, & 0 \end{bmatrix},
\end{aligned}
$$

and

$$
\begin{aligned}
r_D^T &= c_D^T - \lambda^T D \\
&= \begin{bmatrix} c_2 & c_4 & c_5 \end{bmatrix} - \begin{bmatrix} \frac{-6}{5} & \frac{-3}{5} & 0 \end{bmatrix}
\begin{bmatrix}
1 & 1 & 0 \\
2 & 0 & 1 \\
2 & 0 & 0
\end{bmatrix} \\
&= \begin{bmatrix} -1 & 0 & 0 \end{bmatrix} - \begin{bmatrix} \frac{-12}{5} & \frac{-6}{5} & \frac{-3}{5} \end{bmatrix} \\
&= \begin{bmatrix} 7/5 & 6/5 & 3/5 \end{bmatrix} \\
&= \begin{bmatrix} r_2 & r_4 & r_5 \end{bmatrix}.
\end{aligned}
$$

Note that all reduced cost coefficients are positive. Therefore, the current tableau is optimal and the optimal solution is $x = \begin{bmatrix} \frac{1}{5} & 0 & \frac{8}{5} & 0 & 0 & 4 \end{bmatrix}^T$. The optimal value is $-3 \times \frac{1}{5} + (-3) \times \frac{8}{5} = -\frac{27}{5}$.

Example 6.2. Consider the linear programming problem

$$
\begin{aligned}
\text{minimize} \quad & -4x_1 - 3x_2 - 2x_3 \\
\text{subject to} \quad & 2x_1 - 3x_2 + 2x_3 \le 6, \\
& -x_1 + x_2 + x_3 \le 5, \\
& x_1, x_2, \quad x_3 \ge 0.
\end{aligned}
$$

The augmented matrix is

	a_1	a_2	a_3	a_4	a_5	b
a_4	2	-3	2	1	0	6
a_5	-1	1	1	0	1	5

Since basis matrix $B = \begin{bmatrix} a_4 & a_5 \end{bmatrix}$, therefore basic vector is $x_B = \begin{bmatrix} x_4 \\ x_5 \end{bmatrix} = \begin{bmatrix} 6 \\ 5 \end{bmatrix}$. We have a starting basic feasible solution $x = \begin{bmatrix} 0 \\ 0 \\ 0 \\ 6 \\ 5 \end{bmatrix}$.

We determine

$$c_B^T = \begin{bmatrix} c_4 & c_5 \end{bmatrix}, \qquad c_D^T = \begin{bmatrix} c_1 & c_2 & c_3 \end{bmatrix}$$

and

$$\lambda^T = c_B^T B^{-1} = \begin{bmatrix} 0 & 0 \end{bmatrix} \begin{bmatrix} 1 & 0 \\ 0 & 1 \end{bmatrix} = \begin{bmatrix} 0 & 0 \end{bmatrix}.$$

The revised simplex tableau is

	B^{-1}		b
a_4	1	0	6
a_5	0	1	5

We compute

$$r_D^T = c_D^T - \lambda^T D = \begin{bmatrix} c_1 & c_2 & c_3 \end{bmatrix} - \begin{bmatrix} 0 & 0 \end{bmatrix} \begin{bmatrix} a_1 & a_2 & a_3 \end{bmatrix}$$

$$= \begin{bmatrix} -4 & -3 & -2 \end{bmatrix} - \begin{bmatrix} 0 & 0 \end{bmatrix} \begin{bmatrix} 2 & -3 & 2 \\ -1 & 1 & 1 \end{bmatrix}$$

$$= \begin{bmatrix} -4 & -3 & -2 \end{bmatrix}$$

$$= \begin{bmatrix} r_1 & r_2 & r_3 \end{bmatrix}.$$

Since $r_1 = -4$ is most negative, therefore we bring a_1 into basis. However, we first calculate

$$y_1 = B^{-1}a_1$$

$$= \begin{bmatrix} 1 & 0 \\ 0 & 1 \end{bmatrix} \begin{bmatrix} 2 \\ -1 \end{bmatrix} = \begin{bmatrix} 2 \\ -1 \end{bmatrix}.$$

We now form the augmented revised tableau

	B^{-1}		b	y_1
a_4	1	0	6	2
a_5	0	1	5	-1

We find pivot row, that is p = arg min$\left\{ \frac{6}{2}, \cancel{\frac{5}{1}} \right\}$ = 1st row. There-fore, a_4 of basis matrix B is replaced by a_1.

	B^{-1}		b	y_1
a_1	1	0	6	$\boxed{2}$
a_5	0	1	5	-1

We apply elementary row operations by pivoting about the first element of the last column of the augmented revised tableau.

$R_1 \to \frac{1}{2}R_1$

	B^{-1}		b	y_1
a_1	$^{1}/_{2}$	0	3	$\boxed{1}$
a_5	0	1	5	-1

$R_2 \to R_2 + R_1$

	B^{-1}		b	y_1
a_1	$^{1}/_{2}$	0	3	1
a_5	$^{1}/_{2}$	1	8	0

We remove the column y_1 to get the revised simplex tableau

	B^{-1}		b
a_1	$^{1}/_{2}$	0	3
a_5	$^{1}/_{2}$	1	8

We again compute

$$\lambda^T = c_B^T B^{-1} = \begin{bmatrix} c_1 & c_5 \end{bmatrix} \begin{bmatrix} ^{1}/_{2} & 0 \\ ^{1}/_{2} & 1 \end{bmatrix} = \begin{bmatrix} -4 & 0 \end{bmatrix} \begin{bmatrix} ^{1}/_{2} & 0 \\ ^{1}/_{2} & 1 \end{bmatrix} = \begin{bmatrix} -2 & 0 \end{bmatrix}$$

and

$$r_D^T = c_D^T - \lambda^T D = \begin{bmatrix} c_2 & c_3 & c_4 \end{bmatrix} - \begin{bmatrix} -2 & 0 \end{bmatrix} \begin{bmatrix} a_2 & a_3 & a_4 \end{bmatrix}$$

$$= \begin{bmatrix} -3 & -2 & 0 \end{bmatrix} - \begin{bmatrix} -2 & 0 \end{bmatrix} \begin{bmatrix} -3 & 2 & 1 \\ 1 & 1 & 0 \end{bmatrix}$$

$$= \begin{bmatrix} -3 & -2 & 0 \end{bmatrix} - \begin{bmatrix} 6 & -4 & -2 \end{bmatrix}$$

$$= \begin{bmatrix} -9 & 2 & 2 \end{bmatrix}$$

$$= \begin{bmatrix} r_2 & r_3 & r_4 \end{bmatrix}.$$

Since $r_2 = -9$ is only negative, therefore a_2 will come into basis. We also calculate

$$y_2 = B^{-1} a_2$$

$$= \begin{bmatrix} 1/2 & 0 \\ 1/2 & 1 \end{bmatrix} \begin{bmatrix} -3 \\ 1 \end{bmatrix} = \begin{bmatrix} -3/2 \\ -1/2 \end{bmatrix}.$$

We form the augmented revised tableau

	B^{-1}		b	y_2
a_1	$1/2$	0	3	$-3/2$
a_5	$1/2$	1	8	$-1/2$

We find p=arg min$\left\{ \dfrac{3}{-3/2}, \dfrac{8}{-1/2} \right\}$ =no row. Since no $y_{iq} \geq 0$, therefore the problem is unbounded.

We can call MATLAB function `rsm.m` to find the optimal solution of the linear programming problem. See the following `Code 6.1`.

Code 6.1: rsm.m

```
function [Binv,B,xB]=rsm(A,c,B,xB,Binv,v)
%input: augmented matrix A, cost c, basis
%vector B, basic vector xB, identity matrix
%Binv, non-basic variable v
%output: basis vector B, basic vector xB
    [~,n]=size(A);
    format rat
    Cb=c(B);
    Cdd=setdiff(1:n,B);
```

```
D=A( : ,Cdd) ;
Cd=c( 1 ,Cdd) ;
L=Cb∗Binv ;
disp ( 'L= ' ) ;
disp (L) ;
r=Cd−(L∗D) ;
disp ( ' r= ' ) ;
disp ( r ) ;
[ ~ , aa]=pivotcolumn ( r , v ) ;
%check optimality
if aa==0
    disp ( 'optimal solution reached ' ) ;
return
else
    a=Cdd( aa ) ;
    disp ( 'q= ' ) ;
    disp (a) ;
    y=Binv∗A( : , a ) ;
    Y=sprintf ( '%s ' , ' y= ' ) ;
    disp (Y) ;
    disp (y) ;
    revise =[Binv xB y ] ;
    [m, n]= size ( revise ) ;
    min=Inf ;
    arg_min=0;
end
%select pivot row
for k=1:m
    if revise (k ,n)>0
        row=revise (k ,n−1)/ revise (k ,n) ;
        if row<min
            min=row ;
            arg_min=k ;
        end
    end
end
%check unbounded solution
if arg_min==0
    x=sprintf ( '%c ' , 'unbounded ' ) ;
```

```
        disp(x);
    return
    else
        disp('p=');
        disp(arg_min);
    %update revise tableau
        revise=simplex(revise,arg_min,n);
        disp('revise');
        disp(revise);
        Binv=revise(:,1:n-2);
        B(arg_min)=a;
        xB=revise(:,n-1);
    end
end
```

Note that in the above MATLAB function, we have called function **simplex.m** to make the pivot element unity and all values below and up of pivot element in pivot column become zero. We have also called function **pivotcolumn.m** to select pivot column. The MATLAB functions **simplex.m**, **pivotcolumn.m**, and **rsm.m** should be in the same folder in the computer.

Example 6.3. Solve the linear programming problem using the revised simplex method.

$$
\begin{aligned}
\text{maximize} \quad & 3x_1 + 4x_2 + x_3 + 7x_4 \\
\text{subject to} \quad & 8x_1 + 3x_2 + 4x_3 + x_4 \leq 7, \\
& 2x_1 + 6x_2 + x_3 + 5x_4 \leq 3, \\
& x_1 + 4x_2 + 5x_3 + 2x_4 \leq 8, \\
& x_1, \quad x_2, \quad x_3, \quad x_4 \geq 0.
\end{aligned}
$$

Using slack variables x_5, x_6 and x_7 to express the linear programming problem in standard form:

$$
\begin{aligned}
\text{minimize} \quad & -3x_1 - 4x_2 - x_3 - 7x_4 \\
\text{subject to} \quad & 8x_1 + 3x_2 + 4x_3 + x_4 + x_5 && = 7, \\
& 2x_1 + 6x_2 + x_3 + 5x_4 && +x_6 && = 3, \\
& x_1 + 4x_2 + 5x_3 + 2x_4 && +x_7 = 8, \\
& x_1, \quad x_2, \quad x_3, \quad x_4, \quad x_5, x_6, x_7 \geq 0.
\end{aligned}
$$

The augmented matrix is

	a_1	a_2	a_3	a_4	a_5	a_6	a_7	b
a_5	8	3	4	1	1	0	0	7
a_6	2	6	1	5	0	1	0	3
a_7	1	4	5	2	0	0	1	8

The revised simplex tableau is

$$
\begin{array}{c|ccc|c}
 & \multicolumn{3}{c|}{B^{-1}} & \text{b} \\
a_5 & 1 & 0 & 0 & 7 \\
a_6 & 0 & 1 & 0 & 3 \\
a_7 & 0 & 0 & 1 & 8 \\
\end{array}
$$

We compute

$$
\lambda^T = c_B^T B^{-1}
$$

$$
= \begin{bmatrix} c_5 & c_6 & c_7 \end{bmatrix} B^{-1} = \begin{bmatrix} 0 & 0 & 0 \end{bmatrix} \begin{bmatrix} 1 & 0 & 0 \\ 0 & 1 & 0 \\ 0 & 0 & 1 \end{bmatrix} = \begin{bmatrix} 0 & 0 & 0 \end{bmatrix}
$$

and

$$
\begin{aligned}
r_D^T = c_D^T - \lambda^T D &= \begin{bmatrix} c_1 & c_2 & c_3 & c_4 \end{bmatrix} - \begin{bmatrix} 0 & 0 & 0 \end{bmatrix} \begin{bmatrix} 8 & 3 & 4 & 1 \\ 2 & 6 & 1 & 5 \\ 1 & 4 & 5 & 2 \end{bmatrix} \\
&= \begin{bmatrix} -3 & -4 & -1 & -7 \end{bmatrix} - \begin{bmatrix} 0 & 0 & 0 & 0 \end{bmatrix} \\
&= \begin{bmatrix} -3 & -4 & -1 & -7 \end{bmatrix} \\
&= \begin{bmatrix} r_1 & r_2 & r_3 & r_4 \end{bmatrix}.
\end{aligned}
$$

r_4 is the most negative. Therefore, a_4 will enter into basis.

We calculate

$$y = B^{-1}a_4 = \begin{bmatrix} 1 & 0 & 0 \\ 0 & 1 & 0 \\ 0 & 0 & 1 \end{bmatrix} \begin{bmatrix} 1 \\ 5 \\ 2 \end{bmatrix} = \begin{bmatrix} 1 \\ 5 \\ 2 \end{bmatrix}.$$

We now form the augmented revised tableau.

		B^{-1}		b	y_4
a_5	1	0	0	7	1
a_6	0	1	0	3	5
a_7	0	0	1	8	2

We compute pivot row p $= \arg\min\left\{\frac{7}{1}, \frac{3}{5}, \frac{8}{2}\right\} = 2^{\text{nd}}$ row. Therefore, a_6 is replaced by a_4. We call MATLAB functions to perform elementary row operations.

		B^{-1}		b	y_4
a_5	1	0	0	7	1
a_4	0	1	0	3	5
a_7	0	0	1	8	2

$$R_2 \to \tfrac{1}{5}R_2$$

```
>> A = identityop(A,2,1/5)
```

		B^{-1}		b	y_4
a_5	1	0	0	7	1
a_4	0	$1/5$	0	$3/5$	1
a_7	0	0	1	8	2

$$R_1 \to R_1 - R_2$$

```
>> A = eliminationop(A,1,2,-1)
```

		B^{-1}		b	y_4
a_5	1	$-1/5$	0	$32/5$	0
a_4	0	$1/5$	0	$3/5$	1
a_7	0	0	1	8	2

$$R_3 \to R_3 - 2R_2$$

```
>>A=eliminationop(A,3,2,-2)
```

		B^{-1}		b	y_4
a_5	1	$-1/5$	0	$32/5$	0
a_4	0	$1/5$	0	$3/5$	1
a_7	0	$-2/5$	1	$34/5$	0

After removing the last column from the above revised tableau, we get

$$
\begin{array}{c}
\quad B^{-1} \qquad\quad \mathrm{b}\\
\begin{array}{c|ccc|c}
a_5 & 1 & -1/5 & 0 & 32/5\\
a_4 & 0 & 1/5 & 0 & 3/5\\
a_7 & 0 & -2/5 & 1 & 34/5
\end{array}
\end{array}
$$

We compute

$$
\begin{aligned}
\lambda^T &= c_B^T B^{-1}\\
&= \begin{bmatrix} c_5 & c_4 & c_7 \end{bmatrix} B^{-1}\\
&= \begin{bmatrix} 0 & -7 & 0 \end{bmatrix}
\begin{bmatrix}
1 & -1/5 & 0\\
0 & 1/5 & 0\\
0 & -2/5 & 1
\end{bmatrix}\\
&= \begin{bmatrix} 0 & -7/5 & 0 \end{bmatrix}
\end{aligned}
$$

and

$$
\begin{aligned}
r_D^T &= c_D^T - \lambda^T D\\
&= \begin{bmatrix} c_1 & c_2 & c_3 & c_6 \end{bmatrix} - \begin{bmatrix} 0 & -7/5 & 0 \end{bmatrix}
\begin{bmatrix}
8 & 3 & 4 & 0\\
2 & 6 & 1 & 1\\
1 & 4 & 5 & 0
\end{bmatrix}\\
&= \begin{bmatrix} -3 & -4 & -1 & 0 \end{bmatrix} - \begin{bmatrix} -14/5 & -42/5 & -7/5 & -7/5 \end{bmatrix}\\
&= \begin{bmatrix} -3+14/5 & -4+42/5 & -1+7/5 & 0+7/5 \end{bmatrix}\\
&= \begin{bmatrix} -1/5 & 22/5 & 2/5 & 7/5 \end{bmatrix}\\
&= \begin{bmatrix} r_1 & r_2 & r_3 & r_6 \end{bmatrix}.
\end{aligned}
$$

r_1 is the most negative. Therefore, a_1 will enter to basis by

$$
\begin{aligned}
y_1 &= B^{-1} a_1\\
&= \begin{bmatrix}
1 & -1/5 & 0\\
0 & 1/5 & 0\\
0 & -2/5 & 1
\end{bmatrix}
\begin{bmatrix} 8\\ 2\\ 1 \end{bmatrix}
= \begin{bmatrix} 8-2/5\\ 2/5\\ -\frac{4}{5}+1 \end{bmatrix}
= \begin{bmatrix} 38/5\\ 2/5\\ 1/5 \end{bmatrix}.
\end{aligned}
$$

We form the augmented revised tableau.

$$
\begin{array}{c}
\quad B^{-1} \qquad\quad \mathrm{b} \quad y_1\\
\begin{array}{c|ccc|c|c}
a_5 & 1 & -1/5 & 0 & 32/5 & 38/5\\
a_4 & 0 & 1/5 & 0 & 3/5 & 2/5\\
a_7 & 0 & -2/5 & 1 & 34/5 & 1/5
\end{array}
\end{array}
$$

We find pivot row p $= \arg\min\left\{\frac{32}{38}, \frac{3}{2}, 34\right\} = 1^{\text{st}}$ row. Therefore, a_5 is replaced by a_1.

	B^{-1}			b	y_1
a_1	1	$-1/5$	0	$32/5$	$38/5$
a_4	0	$1/5$	0	$3/5$	$2/5$
a_7	0	$-2/5$	1	$34/5$	$1/5$

$$R_1 \to \tfrac{5}{38}R_1$$

```
>>A=identityop(A,1,5/38)
```

	B^{-1}			b	y_1
a_1	$5/38$	$-1/38$	0	$16/19$	1
a_4	0	$1/5$	0	$3/5$	$2/5$
a_7	0	$-2/5$	1	$34/5$	$1/5$

$$R_2 \to R_2 - \tfrac{2}{5}R_1$$

```
>>A=eliminationop(A,2,1,-2/5)
```

	B^{-1}			b	y_1
a_1	$5/38$	$-1/38$	0	$16/19$	1
a_4	$-1/19$	$4/19$	0	$5/19$	0
a_7	0	$-2/5$	1	$34/5$	$1/5$

$$R_3 \to R_3 - \tfrac{1}{5}R_2$$

```
>>A=eliminationop(A,3,1,-1/5)
```

	B^{-1}			b	y_1
a_1	$5/38$	$-1/38$	0	$16/19$	1
a_4	$-1/19$	$4/19$	0	$5/19$	0
a_7	$-1/38$	$-15/38$	1	$126/19$	0

After removing last the column from the above revised tableau, we get

	B^{-1}			b
a_1	$5/38$	$-1/38$	0	$16/19$
a_4	$-1/19$	$4/19$	0	$5/19$
a_7	$-1/38$	$-15/38$	1	$126/19$

We compute

$$\lambda^T = c_B^T B^{-1}$$
$$= \begin{bmatrix} c_1 & c_4 & c_7 \end{bmatrix} B^{-1}$$
$$= \begin{bmatrix} -3 & -7 & 0 \end{bmatrix} \begin{bmatrix} 5/38 & -1/38 & 0 \\ -1/19 & 4/19 & 0 \\ -1/38 & -15/38 & 1 \end{bmatrix}$$
$$= \begin{bmatrix} -\frac{15}{38} + \frac{7}{19} & \frac{3}{38} - \frac{28}{19} & 0 \end{bmatrix}$$
$$= \begin{bmatrix} -\frac{1}{38} & -\frac{53}{38} & 0 \end{bmatrix}.$$

$$r_D^T = c_D^T - \lambda^T D$$

$$= \begin{bmatrix} c_2 & c_3 & c_5 & c_6 \end{bmatrix} - \begin{bmatrix} -\frac{1}{38} & -\frac{53}{38} & 0 \end{bmatrix} \begin{bmatrix} 3 & 4 & 1 & 0 \\ 6 & 1 & 0 & 1 \\ 4 & 5 & 0 & 0 \end{bmatrix}$$
$$= \begin{bmatrix} -4 & -1 & 0 & 0 \end{bmatrix} - \begin{bmatrix} -\frac{321}{38} & -\frac{3}{2} & -\frac{1}{38} & -\frac{53}{38} \end{bmatrix}$$
$$= \begin{bmatrix} \frac{169}{38} & \frac{1}{2} & \frac{1}{38} & \frac{53}{38} \end{bmatrix}$$
$$= \begin{bmatrix} r_2 & r_3 & r_5 & r_6 \end{bmatrix}.$$

All are greater than zero. Therefore, the current revised tableau contains the optimal solution, that is $x_1 = \frac{16}{19}$, $x_4 = \frac{5}{19}$ and the value of an objective function is

$$= -3 \times \frac{16}{19} + -7 \times \frac{5}{19}$$
$$= -\frac{48}{19} - \frac{35}{19}$$
$$= -\frac{83}{19}.$$

Example 6.4. Solve the following linear programming problem using the revised simplex method.

$$\begin{aligned} \text{maximize} \quad & 3x_1 + 5x_2 \\ \text{subject to} \quad & x_1 + x_2 \le 4, \\ & 5x_1 + 3x_2 \ge 8, \\ & x_1, \quad x_2 \ge 0. \end{aligned}$$

In standard form

$$\begin{aligned} \text{minimize} \quad & -3x_1 - 5x_2 \\ \text{subject to} \quad & x_1 + x_2 + x_3 && = 4, \\ & 5x_1 + 3x_2 && -x_4 = 8, \\ & x_1, \quad x_2, \quad x_3, \quad x_4 \geq 0. \end{aligned}$$

There is no obvious basic feasible solution. Therefore, we use the two-phase method.

Phase-I: We introduce an artificial variable

$$\begin{aligned} \text{minimize} \quad & && x_5 \\ \text{subject to} \quad & x_1 + x_2 + x_3 && = 4, \\ & 5x_1 + 3x_2 && -x_4 + x_5 = 8, \\ & x_1, \quad x_2, \quad x_3, \quad x_4, \quad x_5, \geq 0. \end{aligned}$$

The augmented matrix is

a_1	a_2	a_3	a_4	a_5	b
1	1	1	0	0	4
5	3	0	−1	1	8

Note that $B = \begin{bmatrix} a_3 & a_5 \end{bmatrix}$, $x_B = \begin{bmatrix} x_3 \\ x_5 \end{bmatrix} = \begin{bmatrix} 4 \\ 8 \end{bmatrix}$. We have a basic feasible solution. We can start the revised simplex method.

$$\begin{array}{c|cc|c} & \multicolumn{2}{c|}{B^{-1}} & b \\ a_3 & 1 & 0 & 4 \\ a_5 & 0 & 1 & 8 \end{array}$$

We compute

$$\lambda^T = c_B^T B^{-1} = \begin{bmatrix} c_3 & c_5 \end{bmatrix} B^{-1} = \begin{bmatrix} 0 & 1 \end{bmatrix} \begin{bmatrix} 1 & 0 \\ 0 & 1 \end{bmatrix}$$
$$= \begin{bmatrix} 0 & 1 \end{bmatrix}.$$

Therefore,

$$r_D^T = c_D^T - \lambda^T D$$

$$= \begin{bmatrix} c_1 & c_2 & c_4 \end{bmatrix} - \begin{bmatrix} 0 & 1 \end{bmatrix} \begin{bmatrix} 1 & 1 & 0 \\ 5 & 3 & -1 \end{bmatrix}$$

$$= \begin{bmatrix} 0 & 0 & 0 \end{bmatrix} - \begin{bmatrix} 5 & 3 & -1 \end{bmatrix}$$

$$= \begin{bmatrix} -5 & -3 & 1 \end{bmatrix}$$

$$= \begin{bmatrix} r_1 & r_2 & r_4 \end{bmatrix}.$$

The most reduced cost coefficient is r_1. Therefore, a_1 will enter into basis. We calculate

$$y_1 = B^{-1} a_1$$

$$= \begin{bmatrix} 1 & 0 \\ 0 & 1 \end{bmatrix} \begin{bmatrix} 1 \\ 5 \end{bmatrix}$$

$$= \begin{bmatrix} 1 \\ 5 \end{bmatrix}.$$

We write the augmented revised tableau.

	B^{-1}		b	y_1
a_3	1	0	4	1
a_5	0	1	8	5

We find p $=$arg min$\{\frac{4}{1}, \frac{8}{5}\}$=2nd row. Therefore, a_5 is replaced by a_1. We now apply row operations to update the above revised simplex tableau.

	B^{-1}		b	y_1
a_3	1	0	4	1
a_1	0	1	8	$\boxed{5}$

$$R_2 \to \tfrac{1}{5} R_2$$

```
>>A=identityop(A,2,1/5)
```

	B^{-1}		b	y_1
a_3	1	0	4	1
a_1	0	1/5	8/5	$\boxed{1}$

$$R_1 \to R_1 - R_2$$

```
>>A=eliminationop(A,1,2,-1)
```

$$
\begin{array}{c c c c c}
 & B^{-1} & & \mathrm{b} & y_1 \\
a_3 & 1 & -1/5 & 12/5 & 0 \\
a_1 & 0 & 1/5 & 8/5 & \boxed{1}
\end{array}
$$

After removing the last column from the above revised tableau, we get

$$
\begin{array}{c c c c}
 & B^{-1} & & \mathrm{b} \\
a_3 & 1 & -1/5 & 12/5 \\
a_1 & 0 & 1/5 & 8/5
\end{array}
$$

We now compute

$$
\lambda^T = c_B^T B^{-1} = \begin{bmatrix} 0 & 0 \end{bmatrix} B^{-1} = \begin{bmatrix} 0 & 0 \end{bmatrix}
$$

and

$$
r_D^T = c_D^T - \lambda^T D = \begin{bmatrix} c_2 & c_4 & c_5 \end{bmatrix} - \begin{bmatrix} 0 & 0 \end{bmatrix} D = \begin{bmatrix} 0 & 0 & 1 \end{bmatrix}.
$$

All reduced cost coefficients are greater than or equal to zero. Therefore, we stop to get a starting basic feasible solution $x =$
$$
\begin{bmatrix} 8/5 \\ 0 \\ 12/5 \\ 0 \\ 0 \end{bmatrix}
$$ for Phase-II

Phase-II: We now take original objective function for calculating λ^T and r_D^T

$$
\begin{array}{c c c c}
 & B^{-1} & & b \\
a_3 & 1 & -1/5 & 12/5 \\
a_1 & 0 & 1/5 & 8/5
\end{array}
$$

$$
\lambda^T = c_B^T B^{-1} = \begin{bmatrix} c_3 & c_1 \end{bmatrix} \begin{bmatrix} 1 & -1/5 \\ 0 & 1/5 \end{bmatrix} = \begin{bmatrix} 0 & -3 \end{bmatrix} \begin{bmatrix} 1 & -1/5 \\ 0 & 1/5 \end{bmatrix}
$$
$$
= \begin{bmatrix} 0 & -3/5 \end{bmatrix}
$$

and

$$r_D^T = c_D^T - \lambda^T D = \begin{bmatrix} c_2 & c_4 \end{bmatrix} - \begin{bmatrix} 0 & -3/5 \end{bmatrix} \begin{bmatrix} 1 & 0 \\ 3 & -1 \end{bmatrix}$$
$$= \begin{bmatrix} -5 & 0 \end{bmatrix} - \begin{bmatrix} -\frac{9}{5} & \frac{3}{5} \end{bmatrix}$$
$$= \begin{bmatrix} -\frac{16}{5} & -\frac{3}{5} \end{bmatrix}$$
$$= \begin{bmatrix} r_2 & r_4 \end{bmatrix}.$$

Since r_2 is the most negative, therefore a_2 will enter into basis by $y_2 = B^{-1}a_2$. We find

$$y_2 = \begin{bmatrix} 1 & -1/5 \\ 0 & 1/5 \end{bmatrix} \begin{bmatrix} 1 \\ 3 \end{bmatrix} = \begin{bmatrix} 2/5 \\ 3/5 \end{bmatrix}.$$

We form the augmented revised tableau.

	B^{-1}		b	y_2
a_3	1	-1/5	12/5	2/5
a_1	0	1/5	8/5	3/5

We find p=arg min$\{6, \frac{8}{3}\}$=2nd row. Therefore, a_1 is replaced by a_2 We now update the revised simplex tableau.

	B^{-1}		b	y_2
a_3	1	-1/5	12/5	2/5
a_2	0	1/5	8/5	$\boxed{3/5}$

$$R_2 \to \tfrac{5}{3}R_2$$

```
>> A=identityop(A,2,5/3)
```

	B^{-1}		b	y_2
a_3	1	-1/5	12/5	2/5
a_2	0	1/3	8/3	$\boxed{1}$

$$R_1 \to R_1 - \tfrac{2}{5}R_2$$

```
>>A=eliminationop(A,1,2,-2/5)
```

	B^{-1}		b	y_2
a_3	1	-1/3	4/3	0
a_2	0	1/3	8/3	$\boxed{1}$

After removing the last column from the above revised tableau, we get

$$
\begin{array}{ccc}
 & B^{-1} & b \\
a_3 & 1 \quad -1/3 & 4/3 \\
a_2 & 0 \quad 1/3 & 8/3
\end{array}
$$

$$
\lambda^T = c_B^T B^{-1} = \begin{bmatrix} c_3 & c_2 \end{bmatrix} \begin{bmatrix} 1 & -1/3 \\ 0 & 1/3 \end{bmatrix} = \begin{bmatrix} 0 & -5 \end{bmatrix} \begin{bmatrix} 1 & -1/3 \\ 0 & 1/3 \end{bmatrix}
$$
$$
= \begin{bmatrix} 0 & -5/3 \end{bmatrix}.
$$

$$
r_D^T = c_D^T - \lambda^T D = \begin{bmatrix} c_1 & c_4 \end{bmatrix} - \begin{bmatrix} 0 & -5/3 \end{bmatrix} \begin{bmatrix} 1 & 0 \\ 5 & -1 \end{bmatrix}
$$
$$
= \begin{bmatrix} -3 & 0 \end{bmatrix} - \begin{bmatrix} -\frac{25}{3} & \frac{5}{3} \end{bmatrix} = \begin{bmatrix} \frac{16}{3} & -\frac{5}{3} \end{bmatrix}.
$$

r_4 is the most negative. Therefore, a_4 will enter into basis by $y_4 = B^{-1}a_4$.

$$
y_4 = \begin{bmatrix} 1 & -1/3 \\ 0 & 1/3 \end{bmatrix} \begin{bmatrix} 0 \\ -1 \end{bmatrix} = \begin{bmatrix} 1/3 \\ -1/3 \end{bmatrix}.
$$

$$
\begin{array}{cccc}
 & B^{-1} & b & y_4 \\
a_3 & 1 \quad -1/3 & 4/3 & 1/3 \\
a_2 & 0 \quad 1/3 & 8/3 & -1/3
\end{array}
$$

We find p $=$arg min$\{4, \cancel{8}\}=1^{\text{st}}$ row. Therefore, a_3 is replaced by a_4. We now update the revised simplex tableau.

$$
\begin{array}{cccc}
 & B^{-1} & b & y_2 \\
a_4 & 1 \quad -1/3 & 4/3 & \boxed{1/3} \\
a_2 & 0 \quad 1/3 & 8/3 & -1/3 \\
 & R_1 \to 3R_1
\end{array}
$$

```
>> A = identityop(A,1,3)
```

$$
\begin{array}{cccc}
 & B^{-1} & b & y_2 \\
a_4 & 3 \quad -1 & 4 & \boxed{1} \\
a_2 & 0 \quad 1/3 & 8/3 & -1/3 \\
 & R_2 \to R_2 + \frac{1}{3}R_1
\end{array}
$$

```
>> A = eliminationop(A,2,1,1/3)
```

$$
\begin{array}{cccc}
 & B^{-1} & b & y_2 \\
a_4 & 3 \quad -1 & 4 & \boxed{1} \\
a_2 & 1 \quad 0 & 4 & 0
\end{array}
$$

after removing the last column from the above revised tableau, we get

$$
\begin{array}{ccc}
 & B^{-1} & b \\
a_4 & 3 \ \ -1 & 4 \\
a_2 & 1 \ \ \ \ 0 & 4
\end{array}
$$

$$
\lambda^T = c_B^T B^{-1} = \begin{bmatrix} c_4 & c_2 \end{bmatrix} \begin{bmatrix} 3 & -1 \\ 1 & 0 \end{bmatrix} = \begin{bmatrix} 0 & -5 \end{bmatrix} \begin{bmatrix} 3 & -1 \\ 1 & 0 \end{bmatrix} = \begin{bmatrix} -5 & 0 \end{bmatrix}
$$

and

$$
r_D^T = c_D^T - \lambda^T D = \begin{bmatrix} c_1 & c_3 \end{bmatrix} - \begin{bmatrix} -5 & 0 \end{bmatrix} \begin{bmatrix} 1 & 1 \\ 5 & 0 \end{bmatrix}
$$
$$
= \begin{bmatrix} -3 & 0 \end{bmatrix} + \begin{bmatrix} 5 & 5 \end{bmatrix} = \begin{bmatrix} 2 & 5 \end{bmatrix}.
$$

All reduced cost coefficients are nonnegative. Therefore, $x = \begin{bmatrix} 0 \\ 4 \\ 0 \\ 4 \end{bmatrix}$.

The optimal solution to the original problem is $x_1 = 0$, $x_2 = 4$ and the value of the objective function is $3x_1 + 5x_2 = 3 \times 0 + 5 \times 4 = 20$.

Example 6.5. Solve the following problem by the revised simplex method.

$$
\begin{array}{rl}
\text{maximize} & 6x_1 - 2x_2 + 3x_3 \\
\text{subject to} & 2x_1 - x_2 + 2x_3 \le 2, \\
& x_1 \qquad\qquad 4x_3 \le 4, \\
& x_1, \quad x_2, \quad x_3, \ge 0.
\end{array}
$$

We can call MATLAB function `rsm.m` to solve the above problem in MATLAB.

In the Command Window,

```
>> A = [2 -1 2 1 0; 1 0 4 0 1]
>> c = [-6 2 -3 0 0]
>> Binv = eye(2)
>> v = 3
>> B = [4 5]
```

```
>> xB = [2;4]
>> [Binv,B,xB] = rsm(A,c,B,xB,Binv,v)
```

L=

  0   0

r=

  −6   2   −3

q=

  1

y=

  2
  1

p=

  1

revise=

  1/2   0   1   1
  −1/2   1   3   0

Binv =

  1/2   0
  −1/2   1

 B =

  1   5

 xB =

  1
  3

```
>> [Binv,B,xB] = rsm(A,c,B,xB,Binv,v)
```

L=

  −3   0

r=

  −1   3   3

q=

  2

y=

  −1/2
  1/2

p=

  2

revise=

$$
\begin{array}{cccc}
0 & 1 & 4 & 0 \\
-1 & 2 & 6 & 1
\end{array}
$$

Binv =

$$
\begin{array}{cc}
0 & 1 \\
-1 & 2
\end{array}
$$

B =

$$
\begin{array}{cc}
1 & 2
\end{array}
$$

xB =

$$
\begin{array}{c}
4 \\
6
\end{array}
$$

```
>> [Binv,B,b] = rsm(A,c,B,b,Binv,v)
```

L=

$$
\begin{array}{cc}
-2 & -2
\end{array}
$$

r=

$$
\begin{array}{ccc}
9 & 2 & 2
\end{array}
$$

optimal solution reached

Binv =

$$
\begin{array}{cc}
0 & 1 \\
-1 & 2
\end{array}
$$

B =

$$
\begin{array}{cc}
1 & 2
\end{array}
$$

xB =

$$
\begin{array}{c}
4 \\
6
\end{array}
$$

We get $x_1 = 4$, $x_2 = 6$ and objective value is $6 \times 4 - 2 \times 6 = 12$.

Example 6.6. [7]A multistage stratified random sampling technique was adopted for the selection of one block (Ghosi), five villages and 60 farmers in the Mau district of Uttar Pradesh, India in 2003. Out of 71 districts, the Mau district had been purposely selected because the maximum area of the district was covered under agricultural practices and the majority of the farmers were mostly dependent on traditional agricultural practices, which ultimately resulted in a lower per capita district. For this reason, the Mau district was selected.

The basic objective was to develop an optimal resource plan for the farmers of the Mau district so that they could optimize their income and employment level under the given constraints.

maximize $\quad 5334x_{01} + 4670x_{02} + 3701x_{03} + 6805x_{04} + 16387x_{05}$
$\quad\quad + 5024x_{06} + 4193x_{07} + 5572x_{08} + 18907x_{09} + 12165x_{10}$
$\quad\quad + 7176x_{11} + 8732x_{12} - 40x_{13} - 40x_{14} - 0.14x_{15} - 0.14x_{16}$

subject to $\quad x_{01}+x_{02}+x_{03}+x_{04}+x_{05} \quad\quad\quad\quad\quad \le 3.44,$
$\quad\quad\quad x_{04}+x_{05}+x_{06}+x_{07}+x_{08}+x_{09}+x_{10}+x_{11}+x_{12} \le 3.44,$
$\quad\quad\quad 78x_{01} + 71x_{02} + 36x_{03} + 46x_{04} + 45x_{05} \quad\quad\quad \le 661,$
$\quad\quad\quad 16x_{04} + 98x_{05} + 66x_{06} + 57x_{07} + 54x_{08} + 88x_{09} + 41x_{11}$
$\quad\quad\quad\quad\quad\quad\quad\quad\quad\quad\quad\quad\quad\quad\quad + 36x_{12} \le 661,$
$\quad\quad 4125x_{01}+4021x_{02}+2018x_{03}+2898x_{04}+4335x_{05} \le 5800,$
$\quad\quad 638x_{04}+7548x_{05}+4368x_{06}+3812x_{07}+3905x_{08}$
$\quad\quad\quad\quad + 12497x_{09} + 3990x_{10} + 3841x_{11} + 3277x_{12} \le 5800,$
$\quad\quad\quad\quad\quad\quad\quad$ where $x_i \ge 0$ for all $i = 1, \ldots, 12.$

The augmented matrix A is:

a_{01}	a_{02}	a_{03}	a_{04}	a_{05}	a_{06}	a_{07}	a_{08}	a_{09}	a_{10}
1	1	1	1	1	0	0	0	0	0
0	0	0	1	1	1	1	1	1	1
78	71	36	46	45	0	0	0	0	0
0	0	0	16	98	66	57	54	88	33
4125	4021	2018	2898	4335	0	0	0	0	0
0	0	0	638	7548	4368	3812	3905	12497	3990

a_{11}	a_{12}	a_{13}	a_{14}	a_{15}	a_{16}	a_{17}	a_{18}	a_{19}	a_{20}	a_{21}	a_{22}
0	0	0	0	0	0	1	0	0	0	0	0
1	1	0	0	0	0	0	1	0	0	0	0
0	0	0	0	0	0	0	0	1	0	0	0
41	36	0	0	0	0	0	0	0	1	0	0
0	0	0	0	0	0	0	0	0	0	1	0
3841	3277	0	0	0	0	0	0	0	0	0	1

We have basis matrix B=$\begin{bmatrix} a_{17} & a_{18} & a_{19} & a_{20} & a_{21} & a_{22} \end{bmatrix}$.

The starting basic feasible solution is given as

$$x = [0 \quad 0 \quad 0 \quad 0 \quad 0 \quad 0 \quad 0 \quad 0 \quad 0 \quad 0 \quad 0 \quad 0 \quad 0 \quad 0$$
$$0 \quad 0 \quad 3.44 \quad 3.44 \quad 661 \quad 661 \quad 5800 \quad 5800]^{\mathrm{T}}.$$

We form a revised tableau corresponding to a starting basic feasible solution.

	B^{-1}						b
a_{17}	1	0	0	0	0	0	3.44
a_{18}	0	1	0	0	0	0	3.44
a_{19}	0	0	1	0	0	0	661
a_{20}	0	0	0	1	0	0	661
a_{21}	0	0	0	0	1	0	5800
a_{22}	0	0	0	0	0	1	5800

We call MATLAB function **rsm.m** to solve this problem in the MATLAB. Therefore, we take the value of all arguments used in this function. These are

In the Command Window,

```
>> A=[ 1 1 1 1 1 0 0 0 0 0 0 0 0 0 0 0 0 1 0 0 0 0 0; 0
0 0 1 1 1 1 1 1 1 1 1 0 0 0 0 0 1 0 0 0 0; 78 71 36
46 45 0 0 0 0 0 0 0 0 0 0 0 0 0 1 0 0 0;0 0 0 16 98 66
57 54 88 33 41 36 0 0 0 0 0 0 0 1 0 0 ; 4125 4021 2018
2898 4335 0 0 0 0 0 0 0 0 0 0 0 0 0 0 1 0; 0 0 0 638
7548 4368 3812 3905 12497 3990 3841 3277 0 0 0 0 0 0 0
0 0 1]

>> c = [-5334 -4670 -3701 -6805 -16387 -5024 -4193
-5572 -18907 -12165 -7176 -8732 40 40 0.14 0.14 0 0 0
0 0 0]

>> Binv = eye(6)

>> v = 16

>> xB = [3.44;3.44;661;661;5800;5800]
```

```
>> B = [17 18 19 20 21 22]
```

In the Command Window,

```
>> [Binv,B,xB] = rsm(A, c, B, xB, Binv, v)
```

OUTPUT:

$$L = \begin{bmatrix} 0 & 0 & 0 & 0 & 0 & 0 \end{bmatrix}$$

$$r = [\begin{matrix} -5334 & -4670 & -3701 & -6805 & -16387 & -5024 & -4193 \end{matrix}$$
$$\begin{matrix} -5572 & -18907 & -12165 & -7176 & -8732 & 40 & 40 & 7/50 & 7/50 \end{matrix}]$$

q=9

$$y = \begin{bmatrix} 9 \\ 1 \\ 0 \\ 88 \\ 0 \\ 12497 \end{bmatrix} \quad p=6$$

$$\text{revise} = \begin{bmatrix} 1 & 0 & 0 & 0 & 0 & 0 & 86/25 & 0 \\ 0 & 1 & 0 & 0 & 0 & -1/12497 & 2345/788 & 0 \\ 0 & 0 & 1 & 0 & 0 & 0 & 661 & 0 \\ 0 & 0 & 0 & 1 & 0 & -88/12497 & 11783/19 & 0 \\ 0 & 0 & 0 & 0 & 1 & 0 & 5800 & 0 \\ 0 & 0 & 0 & 0 & 0 & 1/12497 & 666/1435 & 1 \end{bmatrix}$$

$$\text{Binv} = \begin{bmatrix} 1 & 0 & 0 & 0 & 0 & 0 \\ 0 & 1 & 0 & 0 & 0 & -1/12497 \\ 0 & 0 & 1 & 0 & 0 & 0 \\ 0 & 0 & 0 & 1 & 0 & -88/12497 \\ 0 & 0 & 0 & 0 & 1 & 0 \\ 0 & 0 & 0 & 0 & 0 & 1/12497 \end{bmatrix}$$

$$B = \begin{bmatrix} 17 & 18 & 19 & 20 & 21 & 9 \end{bmatrix}$$

$$xB = \begin{bmatrix} 86/25 \\ 2345/788 \\ 661 \\ 11783/19 \\ 5800 \\ 666/1435 \end{bmatrix}$$

In the Command Window,

```
>> [Binv,B,xB] = rsm(A, c, B, xB, Binv, v)
```

OUTPUT:

$$L = \begin{bmatrix} 0 & 0 & 0 & 0 & 0 & -761/503 \end{bmatrix}$$

$$r = [-5334 \quad -4670 \quad -3701 \quad -23359/4 \quad -54642/11 \quad 45949/29 \quad 29911/19$$
$$28557/85 \quad -98055/16 \quad -39581/29 \quad -75483/20 \quad 40 \quad 40 \quad 7/50 \quad 7/50 \quad 761/503]$$

q=10

$$y = \begin{bmatrix} 0 \\ 113/166 \\ 0 \\ 1476/301 \\ 0 \\ 3990/12497 \end{bmatrix} \quad p=6$$

$$revise = \begin{bmatrix} 1 & 0 & 0 & 0 & 0 & 0 & 86/25 & 0 \\ 0 & 1 & 0 & 0 & 0 & -1/3990 & 3788/1907 & 0 \\ 0 & 0 & 1 & 0 & 0 & 0 & 661 & 0 \\ 0 & 0 & 0 & 1 & 0 & -11/1330 & 20230/33 & 0 \\ 0 & 0 & 0 & 0 & 1 & 0 & 5800 & 0 \\ 0 & 0 & 0 & 0 & 0 & 1/3990 & 580/399 & 1 \end{bmatrix}$$

$$Binv = \begin{bmatrix} 1 & 0 & 0 & 0 & 0 & 0 \\ 0 & 1 & 0 & 0 & 0 & -1/3990 \\ 0 & 0 & 1 & 0 & 0 & 0 \\ 0 & 0 & 0 & 1 & 0 & -11/1330 \\ 0 & 0 & 0 & 0 & 1 & 0 \\ 0 & 0 & 0 & 0 & 0 & 1/3990 \end{bmatrix}$$

$B = \begin{bmatrix} 17 & 18 & 19 & 20 & 21 & 10 \end{bmatrix}$

$$xB = \begin{bmatrix} 86/25 \\ 3788/1907 \\ 661 \\ 20230/33 \\ 5800 \\ 580/399 \end{bmatrix}$$

In the Command Window,

```
>> [Binv,B,xB]=rsm(A,c,B,xB,Binv,v)
```

OUTPUT:

$L = \begin{bmatrix} 0 & 0 & 0 & 0 & 0 & -811/266 \end{bmatrix}$

$r = [\begin{matrix} -5334 & -4670 & -3701 & -53458/11 & 59633/9 & 157576/19 & 74293/10 \end{matrix}$
$\quad\quad \begin{matrix} 82340/13 & 76779/4 & 31743/7 & 16369/13 & 40 & 40 & 7/50 & 7/50 \end{matrix}] \quad 811/266]$

q=1

$$y = \begin{bmatrix} 1 \\ 0 \\ 78 \\ 0 \\ 4125 \\ 0 \end{bmatrix}$$

p=5

$$revise = \begin{bmatrix} 1 & 0 & 0 & 0 & -1/4125 & 0 & 1678/825 & 0 \\ 0 & 1 & 0 & 0 & 0 & -1/3990 & 3788/1907 & 0 \\ 0 & 0 & 1 & 0 & -26/1375 & 0 & 30323/55 & 0 \\ 0 & 0 & 0 & 1 & 0 & -11/1330 & 20230/33 & 0 \\ 0 & 0 & 0 & 0 & 1/4125 & 0 & 232/165 & 1 \\ 0 & 0 & 0 & 0 & 0 & 1/3990 & 580/399 & 0 \end{bmatrix}$$

$$
\text{Binv} =
\begin{bmatrix}
1 & 0 & 0 & 0 & -1/4125 & 0 \\
0 & 1 & 0 & 0 & 0 & -1/3990 \\
0 & 0 & 1 & 0 & -26/1375 & 0 \\
0 & 0 & 0 & 1 & 0 & -11/1330 \\
0 & 0 & 0 & 0 & 1/4125 & 0 \\
0 & 0 & 0 & 0 & 0 & 1/3990
\end{bmatrix}
$$

$$
\text{B} = \begin{bmatrix} 17 & 18 & 19 & 20 & 1 & 10 \end{bmatrix}
$$

$$
\text{xB} =
\begin{bmatrix}
1678/825 \\
3788/1907 \\
30323/55 \\
20230/33 \\
232/165 \\
580/399
\end{bmatrix}
$$

In the Command Window,

```
>> [Binv,B,xB]=rsm(A,c,B,xB,Binv,v)
```

OUTPUT:

$$
L = \begin{bmatrix} 0 & 0 & 0 & 0 & -1778/1375 & -811/266 \end{bmatrix}
$$

$$
\begin{aligned}
r = \big[& 14297/27 \quad -38204/35 \quad -47835/43 \quad -85620/7 \quad 157576/19 \quad 74293/10 \\
& 82340/13 \quad 76779/4 \quad 31743/7 \quad 16369/13 \quad 40 \quad 40 \quad 7/50 \quad 7/50 \quad 1778/1375 \\
& \qquad\qquad\qquad\qquad\qquad\qquad\qquad\qquad\qquad\qquad\qquad\qquad 811/266 \big]
\end{aligned}
$$

q=4

$$
y =
\begin{bmatrix}
409/1375 \\
1676/1995 \\
-3625/412 \\
2209/206 \\
966/1375 \\
319/1995
\end{bmatrix}
$$

p=5

$$\text{revise=}\begin{bmatrix} 1 & 0 & 0 & 0 & -1/2898 & 0 & 1125/782 & 0 \\ 0 & 1 & 0 & 0 & -17/58643 & -1/3990 & 262/859 & 0 \\ 0 & 0 & 1 & 0 & -1/63 & 0 & 35843/63 & 0 \\ 0 & 0 & 0 & 1 & -103/27836 & -11/1330 & 30170/51 & 0 \\ 0 & 0 & 0 & 0 & 1/2898 & 0 & 1451/725 & 1 \\ 0 & 0 & 0 & 0 & -7/126867 & 1/3990 & 806/711 & 0 \end{bmatrix}$$

$$\text{Binv=}\begin{bmatrix} 1 & 0 & 0 & 0 & -1/2898 & 0 \\ 0 & 1 & 0 & 0 & -17/58643 & -1/3990 \\ 0 & 0 & 1 & 0 & -1/63 & 0 \\ 0 & 0 & 0 & 1 & -103/27836 & -11/1330 \\ 0 & 0 & 0 & 0 & 1/2898 & 0 \\ 0 & 0 & 0 & 0 & -7/126867 & 1/3990 \end{bmatrix}$$

$$\text{B=}\begin{bmatrix} 17 & 18 & 19 & 20 & 4 & 10 \end{bmatrix}$$

$$\text{xB=}\begin{bmatrix} 1125/782 \\ 262/859 \\ 35843/63 \\ 30170/51 \\ 1451/725 \\ 806/711 \end{bmatrix}$$

In the Command Window,

```
>> [Binv,B,xB]=rsm(A,c,B,xB,Binv,v)
```

OUTPUT:

$$L = \begin{bmatrix} 0 & 0 & 0 & 0 & -815/486 & -811/266 \end{bmatrix}$$

$$r = \begin{bmatrix} 14251/9 & 49753/24 & -16162/51 & 27791/2 & 157576/19 & 74293/10 & 82340/13 \\ & 76779/4 & 31743/7 & 16369/13 & 40 & 40 & 7/50 & 7/50 & 815/486 & 811/266 \end{bmatrix}$$

q=3

$$\text{y=}\begin{bmatrix} 440/1449 \\ -1084/1853 \\ 250/63 \\ -3517/471 \\ 1009/1449 \\ -529/4751 \end{bmatrix}$$

p=5

$$
\text{revise} =
\begin{bmatrix}
1 & 0 & 0 & 0 & -1/2018 & 0 & 1628/2877 & 0 \\
0 & 1 & 0 & 0 & 0 & -1/3990 & 3788/1907 & 0 \\
0 & 0 & 1 & 0 & -18/1009 & 0 & 17841/32 & 0 \\
0 & 0 & 0 & 1 & 0 & -11/1330 & 20230/33 & 0 \\
0 & 0 & 0 & 0 & 1/2018 & 0 & 2900/1009 & 1 \\
0 & 0 & 0 & 0 & 0 & 1/3990 & 580/399 & 0
\end{bmatrix}
$$

$$
\text{Binv} =
\begin{bmatrix}
1 & 0 & 0 & 0 & -1/2018 & 0 \\
0 & 1 & 0 & 0 & 0 & -1/3990 \\
0 & 0 & 1 & 0 & -18/1009 & 0 \\
0 & 0 & 0 & 1 & 0 & -11/1330 \\
0 & 0 & 0 & 0 & 1/2018 & 0 \\
0 & 0 & 0 & 0 & 0 & 1/3990
\end{bmatrix}
$$

$$
\text{B} = \begin{bmatrix} 17 & 18 & 19 & 20 & 3 & 10 \end{bmatrix}
$$

$$
\text{xB} =
\begin{bmatrix}
1628/2877 \\
3788/1907 \\
17841/32 \\
20230/33 \\
2900/1009 \\
580/399
\end{bmatrix}
$$

In the Command Window,

```
>> [Binv,B,b]=rsm(A,c,B,b,Binv,v)
```

OUTPUT:

$$
L = \begin{bmatrix} 0 & 0 & 0 & 0 & -3701/2018 & -811/266 \end{bmatrix}
$$

$$
r = \begin{bmatrix} 69168/31 & 137929/51 & 9557/21 & 58305/4 & 157576/19 & 74293/10 & 82340/13 \\
76779/4 & 31743/7 & 16369/13 & 40 & 40 & 7/50 & 7/50 & 815/486 & 811/266 \end{bmatrix}
$$

Optimal solution reached.

$$
\text{B} = \begin{bmatrix} 17 & 18 & 19 & 20 & 3 & 10 \end{bmatrix}
$$

$$
\text{xB} =
\begin{bmatrix}
1628/2877 \\
3788/1907 \\
17841/32 \\
20230/33 \\
2900/1009 \\
580/399
\end{bmatrix} .
$$

6.4 Exercises

Exercise 6.1. Solve the linear programming problem.

$$
\begin{aligned}
\text{maximize} \quad & 3x_1 + x_2 + 5x_3 + 4x_4 \\
\text{subject to} \quad & 3x_1 - 3x_2 + 2x_3 + 8x_4 \leq 50, \\
& 4x_1 + 6x_2 - 4x_3 - 4x_4 \leq 40, \\
& 4x_1 - 2x_2 + x_3 + 3x_4 \leq 20, \\
& x_1, \quad x_2, \quad x_3, \quad x_4 \geq 0.
\end{aligned}
$$

Exercise 6.2. Solve the following linear program by the revised simplex method in MATLAB.

$$
\begin{aligned}
\text{maximize} \quad & x_1 - 3x_2 + x_3 \\
\text{subject to} \quad & 2x_1 + x_2 + x_3 \leq 6, \\
& x_1 + x_2 - x_3 \leq 40, \\
& x_1, \quad x_2, \quad x_3 \geq 0.
\end{aligned}
$$

Exercise 6.3. Solve using the revised simplex method.

$$
\begin{aligned}
\text{maximize} \quad & 2x_1 + x_2 + 3x_3 \\
\text{subject to} \quad & x_1 + 2x_2 + x_3 \leq 6, \\
& 2x_1 + x_3 \leq 4, \\
& x_1, \quad x_2, \quad x_3 \geq 0.
\end{aligned}
$$

Exercise 6.4. Solve the problem.

$$
\begin{aligned}
\text{minimize} \quad & -9x_1 - 10x_2 - 15x_3 \\
\text{subject to} \quad & x_1 + 2x_2 + 5x_3 \leq 45, \\
& 2x_1 + 3x_2 + 3x_3 \leq 60, \\
& x_1 + x_2 + 2x_3 \leq 27, \\
& x_1, \quad x_2, \quad x_3 \geq 0.
\end{aligned}
$$

Exercise 6.5. Consider the following linear programming problem.

$$
\begin{aligned}
\text{minimize} \quad & -4x_1 - 3x_2 \\
\text{subject to} \quad & x_1 + 2x_2 \le 8, \\
& -2x_1 + x_2 \le 5, \\
& 5x_1 + 3x_2 \le 16, \\
& x_1, \quad x_2 \ge 0.
\end{aligned}
$$

(a) Solve the linear programming problem using the simplex method.

(b) Solve the linear programming problem using the revised simplex method.

Exercise 6.6. Solve the linear programming problem using the revised simplex method.

$$
\begin{aligned}
\text{maximize} \quad & 3x_1 + 6x_2 + 2x_3 \\
\text{subject to} \quad & 3x_1 + 4x_2 + x_3 \le 20, \\
& x_1 + 3x_2 + 2x_3 \le 10, \\
& x_1 - x_2 \le 3, \\
& x_3 \le 2, \\
& x, \quad x_2, \quad x_3 \ge 0.
\end{aligned}
$$

Exercise 6.7. Solve the linear programming problem using the revised simplex method.

$$
\begin{aligned}
\text{maximize} \quad & x_1 + 8x_2 + 5x_3 \\
\text{subject to} \quad & x_1 + 4x_2 + 5x_3 \le 7, \\
& 3x_1 + 4x_2 \le 18, \\
& 2x_1 + x_2 \le 7, \\
& x_1, \quad x_2, \quad x_3 \ge 0.
\end{aligned}
$$

Exercise 6.8. Solve the linear programming problem using the revised simplex method.

$$
\begin{aligned}
\text{maximize} \quad & 19x_1 + 13x_2 + 12x_3 + 17x_4 \\
\text{subject to} \quad & 3x_1 + 2x_2 + x_3 + 2x_4 \le 225, \\
& x_1 + x_2 + x_3 + x_4 \le 117, \\
& 4x_1 + 3x_2 + 3x_3 + 4x_4 \le 420, \\
& x_1, \quad x_2, \quad x_3, \quad x_4 \ge 0.
\end{aligned}
$$

Exercise 6.9. Solve the following problems using the revised simplex method.

(a)

$$\text{maximize} \quad 2x_1 + 3x_2$$
$$\text{subject to} \quad 2x_1 + 3x_2 \leq 30,$$
$$x_1 + 2x_2 \geq 10,$$
$$x_1, \quad x_2 \geq 0.$$

(b)

$$\text{maximize} \quad 5x_1 + 6x_2$$
$$\text{subject to} \quad x_1 + x_2 \leq 2,$$
$$4x_1 + x_2 \geq 4,$$
$$x_1, \quad x_2 \geq 0.$$

Exercise 6.10. Solve the following linear program using the revised simplex method.

$$\text{minimize} \quad -4x_1 - 3x_2$$
$$\text{subject to} \quad 5x_1 + x_2 \geq 11,$$
$$2x_1 + x_2 \leq 8,$$
$$x_1 + 2x_2 \geq 7,$$
$$x_1, \quad x_2 \geq 0.$$

Exercise 6.11. Solve the problem.

$$\text{maximize} \quad x_1 + 4x_2$$
$$\text{subject to} \quad 2x_1 + x_2 \leq 7,$$
$$2x_1 + 3x_2 \geq 6,$$
$$2x_1 + 6x_2 \geq 9,$$
$$x_1, \quad x_2 \geq 0.$$

Chapter 7

Duality

7.1 Dual Linear Programs

For every linear programming problem, there is a corresponding dual linear programming problem. The dual linear programming problem is constructed from the cost and constraints of the original linear programming problem or their primal linear programming problem. We know that a linear programming problem is solved by the simplex method. Therefore, a dual linear programming problem can also be solved using the simplex method because every dual linear programming problem is a linear programming problem. Duality is used to improve the performance of the simplex algorithm (leading to dual algorithm). It helped to develop nonsimplex algorithms such as Karmarkar's algorithm and Khachiyan's algorithm. Lemke and Beale in 1954 designed a dual version of the simplex method.

Consider the following linear programming problem of the form:

$$
\begin{aligned}
&\text{minimize} && c^T x, \\
&\text{subject to} && Ax \geq b, \\
&\text{where} && x \geq 0, x \in \mathbb{R}^n, A \in \mathbb{R}^{m \times n}, m < n, b \in \mathbb{R}^m.
\end{aligned}
\tag{7.1}
$$

We refer to the above problem as the primal problem. The cost vector c in the primal will move to the constraints in the dual. The vector b on the right-hand side of $Ax \geq b$ given in (7.1) will become part of the cost in the dual. Thus, we define the corresponding dual problem as follows:

$$
\begin{aligned}
&\text{maximize} && \lambda^T b \\
&\text{subject to} && \lambda^T A \leq c^T, \\
&&& \lambda \geq 0, \lambda \in \mathbb{R}^m.
\end{aligned}
\tag{7.2}
$$

Note:

(a) The cost vector $c \in \mathbb{R}^n$ of the primal linear programming problem has moved to the constraints in the dual linear programming.

(b) The vector $b \in \mathbb{R}^m$ of the constraints of the primal linear programming problem becomes the part of the cost of the dual linear programming problem.

(c) The form of duality defined as (7.2) is called the symmetric form of duality.

(d) To construct the dual of an arbitrary linear programming problem, we follow the procedure: we firstly convert the given linear programming problem into a problem of above form (symmetric form) and then construct the dual as above.

(e) The dual of the dual is the original linear programming problem.

Problem (7.2) can be written in symmetric form of duality as follows:
$$\begin{aligned} \text{minimize} \quad & (+\lambda^T)(-b) \\ \text{subject to} \quad & \lambda^T(-A) \geq -c^T, \\ & \lambda \geq 0. \end{aligned} \tag{7.3}$$

Then, the dual of the above problem is

$$\begin{aligned} \text{maximize} \quad & (-c^T x) \\ \text{subject to} \quad & (-A)x \leq -b, \\ & x \geq 0. \end{aligned} \tag{7.4}$$

We can get back to the primal linear programming problem:

$$\begin{aligned} \text{minimize} \quad & c^T x \\ \text{subject to} \quad & Ax \geq b, \\ & x \geq 0. \end{aligned}$$

Thus, the dual of the dual is the primal problem. Such problems are also called "symmetric dual problems". It is a very important class in nonlinear programming problems.

We now consider the standard form of the linear programming problem as

$$\begin{aligned} \text{minimize} \quad & c^T x \\ \text{subject to} \quad & Ax = b, \\ & x \geq 0. \end{aligned} \tag{7.5}$$

This form has an equality constraint. To obtain the dual problem of (7.5), we convert the constraint in symmetric form, i.e., $Ax \geq b$.

We observe that $Ax = b$ is an equality constraint. We first convert this equality constraint into the inequality constraints as

$$Ax \geq b \text{ and } Ax \leq b.$$

That is,

$$Ax \geq b \text{ and } -Ax \geq -b.$$

Therefore, the standard form of the linear programming problem can be written as:

$$\begin{aligned} \text{minimize} \quad & c^T x \\ \text{subject to} \quad & \begin{bmatrix} A \\ -A \end{bmatrix} x \geq \begin{bmatrix} b \\ -b \end{bmatrix}, \\ & x \geq 0. \end{aligned}$$

and its dual is

$$\text{maximize} \quad \lambda^T \begin{bmatrix} b \\ -b \end{bmatrix}$$

$$\begin{aligned} \text{subject to} \quad & \lambda^T \begin{bmatrix} A \\ -A \end{bmatrix} \leq c^T, \\ & \lambda \geq 0. \end{aligned}$$

Asymmetric form of duality can also be referred to as:

$$\begin{aligned} \text{maximize} \quad & \lambda^T b \\ \text{subject to} \quad & \lambda^T A \leq c^T. \end{aligned}$$

Let $\lambda = u - v$, then the above dual problem becomes

$$\begin{aligned} \text{maximize} \quad & (u - v)^T b \\ \text{subject to} \quad & (u - v)^T A \leq c^T, \\ & u, v \geq 0, u, v \in \mathbb{R}^m. \end{aligned}$$

Note that the dual vector λ is not restricted to be non-negative.

Example 7.1. Find the dual of the following linear programming problem.

$$
\begin{array}{rrrr}
\text{minimize} & x_1 - 2x_2 & & \\
\text{subject to} & x_1 - & x_2 \geq & 2, \\
& -x_1 + & x_2 \geq & -1, \\
& x_1, & x_2 \geq & 0.
\end{array}
$$

The dual of the linear programming problem is

$$
\begin{array}{rrrr}
\text{maximize} & 2\lambda_1 - \lambda_2 & & \\
\text{subject to} & \lambda_1 - \lambda_2 \leq & 1, \\
& -\lambda_1 + \lambda_2 \leq & -2, \\
& \lambda_1, & \lambda_2 \geq & 0.
\end{array}
$$

Example 7.2. Write down the dual of the linear programming problem.

$$
\begin{array}{rrrr}
\text{maximize} & 2x_1 + x_2 & & \\
\text{subject to} & x_1 & \leq & 3, \\
& x_2 \leq & 4, \\
& x_1 + x_2 \leq & 10, \\
& x_1, & x_2 \geq & 0.
\end{array}
$$

The dual of the linear programming problem is

$$
\begin{array}{rrrr}
\text{minimize} & 3\lambda_1 + 4\lambda_2 + 10\lambda_3 & & \\
\text{subject to} & \lambda_1 & + & \lambda_3 \geq 2, \\
& \lambda_2 + & \lambda_3 \geq 1, \\
& \lambda_1, & \lambda_2, & \lambda_3 \geq 0.
\end{array}
$$

7.2   Properties of Dual Problems

In this section, we discuss some basic properties of dual problems. We begin with the weak duality theorem.

Theorem 7.1. [Weak Duality Theorem] *Suppose that x and λ are feasible solutions of the primal and the dual linear programming problems, respectively (either in the symmetric or asymmetric form), then*

$$c^T x \geq \lambda^T b.$$

Proof. We first prove this theorem for the symmetric form of duality.

Case-1: (Symmetric Case) Primal linear programming can be written as

$$\begin{array}{ll} \text{minimize} & c^T x \\ \text{subject to} & Ax \geq b, \\ & x \geq 0. \end{array} \tag{7.6}$$

Its dual problem can also be written as

$$\begin{array}{ll} \text{maximize} & \lambda^T b \\ \text{subject to} & \lambda^T A \leq c^T, \\ & \lambda \geq 0. \end{array} \tag{7.7}$$

Since x is a feasible solution to the primal problem, we have

$$Ax \geq b, \tag{7.8}$$
$$x \geq 0. \tag{7.9}$$

Since λ is a feasible solution to the dual problem, we have

$$\lambda^T A \leq c^T, \tag{7.10}$$
$$\lambda \geq 0. \tag{7.11}$$

Premultiply (7.8) by λ^T, we get

$$\lambda^T A x \geq \lambda^T b. \tag{7.12}$$

Premultiply (7.10) by x, we get

$$\lambda^T A x \leq c^T x. \tag{7.13}$$

From (7.12) and (7.13),

$$\lambda^T b \leq \lambda^T A x \leq c^T x, \tag{7.14}$$

that is

$$\lambda^T b \leq c^T x. \tag{7.15}$$

Case-II: Asymmetric Case:
The primal problem in asymmetric form is

$$\begin{array}{ll} \text{minimize} & c^T x \\ \text{subject to} & Ax = b, \\ & x \geq 0. \end{array}$$

Given that x is a feasible solution to the primal problem. That is

$$Ax = b, \tag{7.16}$$
$$x \geq 0. \tag{7.17}$$

λ is a feasible solution to the dual problem (7.2). That is

$$\lambda^T A \leq c^T. \tag{7.18}$$

Postmultiply (7.18) by x,

$$\lambda^T Ax \leq c^T x. \tag{7.19}$$

From (7.16) and (7.19), we get

$$\lambda^T b \leq c^T x.$$

$\square$

Note: From the weak duality theorem, we have noticed that the value of the objective function of the dual problem is always less than or equal to the value of the objective function of the primal problem. That is, the value of the objective function of the dual problem is a lower bound to the value of objective function of the primal problem. This is one important concept of the duality.

Theorem 7.2. *Suppose that x_0 and λ_0 are feasible solutions to the primal and dual problems, respectively (either in symmetric or asymmetric form). If $c^T x_0 = \lambda_0^T b$, then x_0 and λ_0 are optimal solutions to their respective problems.*

Proof. Let x be an arbitrary feasible solution of the primal problem. Since λ_0 is a feasible solution to the dual problem, then by the weak duality theorem

$$c^T x \geq \lambda_0^T b. \tag{7.20}$$

We have

$$c^T x_0 = \lambda_0^T b. \tag{7.21}$$

From (7.20) and (7.21)

$$c^T x \geq c^T x_0. \tag{7.22}$$

Thus, x_0 is an optimal solution to the primal linear programming problem. Similarly, let λ be an arbitrary feasible solution to the dual problem and x_0 is a feasible solution to the primal problem, then by weak duality theorem, we have

$$c^T x_0 \geq \lambda^T b. \tag{7.23}$$

From (7.21) and (7.23), we get

$$\lambda_0^T b \geq \lambda^T b.$$

Thus, λ_0 is an optimal solution of the dual problem. $\square$

Theorem 7.3. *[**Duality Theorem**] If the primal problem (either in symmetric or asymmetric form) has an optimal solution, then the dual and optimal values of their respective objective functions are equal.*

Proof. **Case I** [Asymmetric Case]:
Assume that the primal problem has an optimal solution. Then, by the fundamental theorem of the linear programming problem (5.2), there exists an optimal basic feasible solution of the linear programming problem. Let B be the basis matrix, D the matrix corresponding to the nonbasic variables, c_B^T the basic matrix, c_D^T the non-basic vector, and r_D^T the reduced cost coefficients vector. Thus,

$$r_D^T = c_D^T - c_B^T B^{-1} D \geq 0,$$

that is

$$c_B^T B^{-1} D \le c_D^T.$$

Define

$$\lambda^T = c_B^T B^{-1},$$

then

$$\lambda^T D \le c_D^T.$$

Claim: λ is a feasible solution to the dual problem.

Without loss of generality, we assume that the basic columns are the first m columns of A. Then,

$$\lambda^T A = \lambda^T [B, \quad D] = [\lambda^T B, \quad \lambda^T D] \le [c_B^T, c_D^T] = c^T,$$

that is

$$\lambda^T A \le c^T.$$

Therefore, λ is feasible for the dual problem.

Claim: λ is an optimal feasible solution of the dual problem.

$$\lambda^T b = c_B^T B^{-1} b.$$

Since

$$x_B = B^{-1} b,$$

then

$$\lambda^T b = c_B^T x_B.$$

By weak duality theorem, λ is optimal to the dual problem.

Case II: [Symmetric Case] Recall primal problem (7.1). We convert this problem to standard form using surplus variables, that is

$$\text{minimize} \quad \begin{bmatrix} c^T & 0^T \end{bmatrix} \begin{bmatrix} x \\ y \end{bmatrix}$$

$$\text{subject to} \quad \begin{bmatrix} A & -I \end{bmatrix} \begin{bmatrix} x \\ y \end{bmatrix} = b, \tag{7.24}$$

$$\begin{bmatrix} x \\ y \end{bmatrix} \ge 0.$$

Note that x is optimal for (7.1) if and only if $\left[x^T, \ (Ax-b)^T\right]^T$ is optimal for (7.24). The result of **Case I** also applies to **Case II**. This completes the proof. $\qquad\square$

Theorem 7.4 (Complementary Slackness Condition). *The feasible solutions x and λ to a primal dual pair of problems (either in symmetric or asymmetric form) are optimal if and only if*

1. $(c^T - \lambda^T A)x = 0$, *and*

2. $\lambda^T(Ax - b) = 0$.

Proof. **Case I:** (Asymmetric Case) Recall the primal linear program (7.5).

Suppose that x and λ are optimal solutions of the primal dual pair linear programming problem, then by Theorem (7.2),

$$c^T x = \lambda^T b. \tag{7.25}$$

Since

$$Ax = b,$$

therefore,

$$c^T x = \lambda^T Ax.$$

That is,

$$(c^T - \lambda^T A)x = 0.$$

Condition 1 is done.

Case II: (Symmetric Case) Suppose that x and λ are optimal solutions of the primal dual pair linear programming problem, then

$$c^T x = \lambda^T b.$$

In case of symmetric,

$$Ax \geq b, \ x \geq 0.$$

Since

$$\begin{aligned}
(c^T - \lambda^T A)x &= c^T x - \lambda^T Ax \\
&= \lambda^T b - \lambda^T Ax \\
&= \lambda^T(b - Ax) \\
&\leq 0,
\end{aligned}$$

that is

$$(c^T - \lambda^T A)x \leq 0. \tag{7.26}$$

On the other hand,

$$\lambda^T A \leq c^T \text{ and } x \geq 0.$$

We can have

$$\lambda^T Ax \leq c^T x,$$

that is

$$(c^T - \lambda^T A)x \geq 0. \tag{7.27}$$

From (7.26) and (7.27), we get

$$(c^T - \lambda^T A)x = 0.$$

Condition 1 is done.
Since

$$Ax \geq b \text{ and } \lambda \geq 0,$$

then

$$\lambda^T(Ax - b) \geq 0 \tag{7.28}$$

On the other hand, recall the symmetric form of the dual problem:

$$\lambda^T A \leq c^T.$$

We can have

$$\lambda^T Ax \leq c^T x \text{ where } x \geq 0.$$

Using (7.25) to get

$$\lambda^T Ax \leq \lambda^T b.$$

Thus,

$$\lambda^T(Ax - b) \leq 0. \tag{7.29}$$

From (7.28) and (7.29),

$$\lambda^T(Ax - b) = 0.$$

Condition 2 is done.

Conversely, suppose that

$$(c^T - \lambda^T A)x = 0,$$

then

$$c^T x = \lambda^T Ax. \tag{7.30}$$

Since

$$\lambda^T(Ax - b) = 0,$$

then

$$\lambda^T Ax = \lambda^T b. \tag{7.31}$$

Combining (7.30) and (7.31), we get

$$c^T x = \lambda^T b.$$

Therefore, x and λ are optimal solutions for primal and dual linear programming problems, respectively. $\qquad\square$

7.3 The Dual Simplex Method

We can find the dual of any linear programming problem. We use an implicit technique that involves the simplex algorithm to the dual tableau. This is known as a "dual simplex method". The dual simplex method is used when there is no obvious basic feasible solution to the linear programming problem.

Algorithm

1. Create a canonical augmented matrix.

2. Is $b_{i0} \geq 0$ for all i? If yes, then go to Step 3, otherwise go to Step 4.

3. The current solution is optimal.

4. Select the pth row as most negative element such that $b_{po} < 0$.

5. If $a_{pj} \geq 0$ for all j, stop; the dual problem is unbounded, otherwise compute q=arg $\min_j \left\{ \mid \frac{c_j}{a_{pj}} \mid; a_{pj} < 0 \right\}$ for selecting the qth column.

6. Update the canonical augmented matrix by pivoting about the $(p,q)^{\text{th}}$ element.

Example 7.3. Solve the following linear programming problem using the dual simplex method.

$$
\begin{aligned}
\text{minimize} \quad & x_1 + x_2 + x_3 \\
\text{subject to} \quad & x_1 + x_2 + x_3 \geq 3, \\
& 4x_1 + x_2 + 2x_3 \geq 5, \\
& x_1, \quad x_2, \quad x_3 \geq 0.
\end{aligned}
$$

$$
\begin{aligned}
\text{minimize} \quad & x_1 + x_2 + x_3 \\
\text{subject to} \quad -\ & x_1 - x_2 - x_3 + x_4 \qquad = -3, \\
& -4x_1 - x_2 - 2x_3 \qquad +x_5 = -5, \\
& x_1, \quad x_2, \quad x_3, \quad x_4, \quad x_5 \geq 0.
\end{aligned}
$$

The augmented matrix form can be given as

	a_1	a_2	a_3	a_4	a_5	b
a_4	-1	-1	-1	1	0	-3
a_5	-4	-1	-2	0	1	-5
c^T	1	1	1	0	0	0

Since the basis matrix $B = \begin{bmatrix} a_4 & a_5 \end{bmatrix}$, therefore $x_B = \begin{bmatrix} x_4 \\ x_5 \end{bmatrix} = \begin{bmatrix} -3 \\ -5 \end{bmatrix}$. This is a basic solution, but this is not feasible. We use the dual simplex method for solving the above linear programming problem. We get p =min $\begin{bmatrix} -3 \\ -5 \end{bmatrix}$ = 2nd row as a pivot row. We now compute q =arg min$\left\{ \mid \frac{1}{-4} \mid, \mid \frac{1}{-1} \mid, \mid \frac{1}{-2} \mid, \cancel{\frac{0}{0}}, \cancel{\frac{0}{1}} \right\}$ = 1st column as a pivot column. We apply row operations to update the tableau by pivoting $(2, 1)^{\text{th}}$ element.

	a_1	a_2	a_3	a_4	a_5	b
a_4	-1	-1	-1	1	0	-3
a_1	-4	-1	-2	0	1	-5
c^T	1	1	1	0	0	0

$R_2 \to -\frac{1}{4}R_2$

	a_1	a_2	a_3	a_4	a_5	b
a_4	-1	-1	-1	1	0	-3
a_1	1	$1/4$	$1/2$	0	$-1/4$	$5/4$
c^T	1	1	1	0	0	0

$R_1 \to R_1 + R_2$
$R_3 \to R_3 - R_2$

	a_1	a_2	a_3	a_4	a_5	b
a_4	0	$-3/4$	$-1/2$	1	$-1/4$	$-7/4$
a_1	1	$1/4$	$1/2$	0	$-1/4$	$5/4$
c^T	0	$3/4$	$1/2$	0	$1/4$	$-5/4$

$R_1 \leftrightarrow R_2$

	a_1	a_2	a_3	a_4	a_5	b
a_1	1	$1/4$	$1/2$	0	$-1/4$	$5/4$
a_4	0	$-3/4$	$-1/2$	1	$-1/4$	$-7/4$
c^T	0	$3/4$	$1/2$	0	$1/4$	$-5/4$

Since $B = \begin{bmatrix} a_1 & a_4 \end{bmatrix}$, therefore basic vector $x_B = \begin{bmatrix} x_1 \\ x_4 \end{bmatrix} = \begin{bmatrix} 5/4 \\ -7/4 \end{bmatrix}$.
The solution is basic but not feasible. Therefore, we again apply the dual simplex method. $-\frac{7}{4}$ is the only negative element. Thus, p=2, that is, the 2nd row is chosen as a pivot row. We find q=arg

$\min\left\{\cancel{\tfrac{0}{4}},|-1|,|-1|,\cancel{\tfrac{0}{4}},|-1|\right\} = 2^{\text{nd}}$ column. We update the tableau.

	a_1	a_2	a_3	a_4	a_5	b
a_1	1	$1/4$	$1/2$	0	$-1/4$	$5/4$
a_4	0	$-3/4$	$-1/2$	1	$-1/4$	$-7/4$
c^T	0	$3/4$	$1/2$	0	$1/4$	$-5/4$

$R_2 \rightarrow -\tfrac{4}{3}R_2$

	a_1	a_2	a_3	a_4	a_5	b
a_1	1	$1/4$	$1/2$	0	$-1/4$	$5/4$
a_2	0	1	$2/3$	$-4/3$	$1/3$	$7/3$
c^T	0	$3/4$	$1/2$	0	$1/4$	$-5/4$

$R_1 \rightarrow R_1 - \tfrac{1}{4}R_2$
$R_3 \rightarrow R_3 - \tfrac{3}{4}R_2$

	a_1	a_2	a_3	a_4	a_5	b
a_1	1	0	$1/3$	$1/3$	$-1/3$	$2/3$
a_2	0	1	$2/3$	$-4/3$	$1/3$	$7/3$
c^T	0	0	0	1	0	-3

Since basis $B = \begin{bmatrix} a_1 & a_2 \end{bmatrix}$, therefore $x_B = \begin{bmatrix} 2/3 \\ 7/3 \end{bmatrix}$. This is basic and feasible also. All $b_{i0} \geq 0$. Thus, the current basic feasible solution is

optimal. Therefore, $x = \begin{bmatrix} x_1 \\ x_2 \\ x_3 \\ x_4 \\ x_5 \end{bmatrix} = \begin{bmatrix} 2/3 \\ 7/3 \\ 0 \\ 0 \\ 0 \end{bmatrix}$ and value of the objective

function is 3.

Example 7.4. Consider the linear programming problem

$$
\begin{aligned}
\text{minimize} \quad & x_1 + 2x_2 \\
\text{subject to} \quad & x_1 - 4x_2 \geq 2, \\
& 2x_1 - 2x_2 \geq 7, \\
& x_1 + 3x_2 \geq -2, \\
& x_1, \quad x_2 \geq 0.
\end{aligned}
$$

Solve the above problem using the dual simplex method.
We write the problem in standard form:

$$
\begin{aligned}
\text{minimize} \quad & x_1 + 2x_2 \\
\text{subject to} \quad & -\ x_1 + 4x_2 \leq -2, \\
& -2x_1 + 2x_2 \leq -7, \\
& -\ x_1 - 3x_2 \leq 2, \\
& x_1, \quad x_2 \geq 0.
\end{aligned}
$$

That is,

$$
\begin{aligned}
\text{minimize} \quad & x_1 + 2x_2 \\
\text{subject to} \quad & -\ x_1 + 4x_2 + x_3 \qquad\qquad = -2, \\
& -2x_1 + 2x_2 \qquad +x_4 \qquad = -7, \\
& -\ x_1 - 3x_2 \qquad\qquad +x_5 = 2, \\
& x_1, \quad x_2, \quad x_3, \ x_4, \ x_5 \geq 0.
\end{aligned}
$$

The augmented matrix of the above problem is

	a_1	a_2	a_3	a_4	a_5	b
a_3	-1	4	1	0	0	-2
a_4	-2	2	0	1	0	-7
a_5	-1	-3	0	0	1	2
c^T	1	2	0	0	0	0

Basis is $B = \begin{bmatrix} a_3 & a_4 & a_5 \end{bmatrix}$, therefore basic vector is $x_B = \begin{bmatrix} -2 \\ -7 \\ 2 \end{bmatrix}$. It

has a basic solution, but it is not feasible. We now apply the dual

simplex method. We find p= $\min\{x_B\} = \min \begin{bmatrix} -2 \\ -7 \\ 2 \end{bmatrix} = 2^{\text{nd}}$ row as a pivot row and $q= \arg\min\left\{\,\left|\frac{1}{-2}\right|, \cancel{\frac{2}{2}}, \cancel{\frac{0}{0}}, \cancel{\frac{0}{0}}, \cancel{\frac{0}{0}}\,\right\} = 1^{\text{st}}$ column as a pivot column. We update the tableau.

	a_1	a_2	a_3	a_4	a_5	b
a_3	-1	4	1	0	0	-2
a_1	-2	2	0	1	0	-7
a_5	-1	-3	0	0	1	2
c^T	1	2	0	0	0	0

$R_2 \to -\frac{1}{2}R_2$

	a_1	a_2	a_3	a_4	a_5	b
a_3	-1	4	1	0	0	-2
a_1	1	-1	0	$-1/2$	0	$7/2$
a_5	-1	-3	0	0	1	2
c^T	1	2	0	0	0	0

$R_1 \to R_1 + R_2$
$R_3 \to R_3 + R_2$
$R_4 \to R_4 - R_2$

	a_1	a_2	a_3	a_4	a_5	b
a_3	0	3	1	$-1/2$	0	$3/2$
a_1	1	-1	0	$-1/2$	0	$7/2$
a_5	0	-4	0	$-1/2$	1	$11/2$
c^T	0	3	0	$1/2$	0	$-7/2$

All $b_{i0} \geq 0$, which ultimately leads to an optimal solution for the problem. Thus $x_1 = \frac{7}{2}$, $x_2 = 0$, $x_3 = \frac{3}{2}$ and the value of the objective function is $\frac{7}{2}$.

We can solve the above problems in MATLAB. MATLAB function `dual.m` is written in the following Code 7.1 to choose the pivot row and pivot column leading to the pivot element.

Code 7.1: dual.m

```
function  [A,p,q,e,B]  =  dual(A,B)
%input:  augmented  matrix  A,  basis  matrix  B
%output:  augmented  matrix  A,  pivotrow  p,
%  pivotcolumn  q,  pivotelement  e,
%  basis  matrix  B
[m,n]=size(A);
min=0;
p=0;
for  I=1:m−1
    if  A(I,n)<  min
        min=A(I,n);
        p=I;
    end
end
if  p==0
    disp('optimal  solution  reached');
    e=0;
    q=0;
    return;
end
min  =Inf;
q=0;
count=0;
for  k  =  1:n−1
    if  A(m,k) ˜=0
        if  A(p,k)<0
            col=  abs(A(m,k)/A(p,k));
            if  col  <  min
                min  =  col;
                q  =  k;
            end
        end
    end
end
for  k  =  1:n−1
    if  A(p,k)>=0
        count=count+1;
```

```
        end
    end
    if  count==n−1
        disp('unbounded');
        e=0;
        q=0;
        return
    end
    e=A(p,q);
    B(p)=q;
    return
```

Example 7.5. Solve the following linear programming problem by the dual simplex method:

$$
\begin{aligned}
\text{minimize} \quad & 3x_1 + x_2 \\
\text{subject to} \quad & x_1 + x_2 \geq 1, \\
& 2x_1 + 3x_2 \geq 2, \\
& x_1, \quad x_2 \geq 0.
\end{aligned}
$$

We need to express the problem in standard form:

$$
\begin{aligned}
\text{minimize} \quad & 3x_1 + x_2 \\
\text{subject to} \quad & -x_1 - x_2 \leq -1, \\
& -2x_1 - 3x_2 \leq -2, \\
& x_1, \quad x_2 \geq 0.
\end{aligned}
$$

That is,

$$
\begin{aligned}
\text{minimize} \quad & 3x_1 + x_2 \\
\text{subject to} \quad & -x_1 - x_2 + x_3 \qquad = -1, \\
& -2x_1 - 3x_2 \qquad + x_4 = -2, \\
& x_1, \quad x_2, \quad x_3, \quad x_4 \geq 0.
\end{aligned}
$$

The augmented matrix of above linear equality constraints:

	a_1	a_2	a_3	a_4	b
a_3	−1	−1	1	0	−1
a_4	−2	−3	0	1	−2

Since basis $B = \begin{bmatrix} a_3 & a_4 \end{bmatrix}$. Therefore, $x_B = \begin{bmatrix} x_3 \\ x_4 \end{bmatrix} = \begin{bmatrix} -1 \\ -2 \end{bmatrix}$. This is a basic solution, but it is not feasible. Therefore, for solving such problem, we use the dual simplex method. The augmented matrix form of the above standard problem is

	a_1	a_2	a_3	a_4	b
a_3	-1	-1	1	0	-1
a_4	-2	-3	0	1	-2
c^T	3	1	0	0	0

We use MATLAB function `dual.m` to find the pivot row and column resulting pivot element.

```
>> B = [3  4]
>> [A,p,q,e,B] = dual(A,B)
```

	a_1	a_2	a_3	a_4	b
a_3	-1	-1	1	0	-1
a_4	-2	-3	0	1	-2
c^T	3	1	0	0	0

```
p = 2   q = 2   e = -3   B = [3  2] .
```

We find pivot row $p = \min\{x_B\} = \min\begin{bmatrix} -1 \\ -2 \end{bmatrix} = 2^{\text{nd}}$ row and pivot

column $q = \arg\min\left\{ \mid \frac{3}{-2} \mid, \mid \frac{1}{-3} \mid, \cancel{\frac{0}{0}}, \cancel{\frac{0}{1}} \right\} = 2^{\text{nd}}$ column.

	a_1	a_2	a_3	a_4	b
a_3	-1	-1	1	0	-1
a_2	-2	-3	0	1	-2
c^T	3	1	0	0	0

We call MATLAB functions to perform elementary row operations.

	a_1	a_2	a_3	a_4	b
a_3	-1	-1	1	0	-1
a_2	-2	-3	0	1	-2
c^T	3	1	0	0	0

$$R_2 \to -\tfrac{1}{3} R_2$$

```
>> A = identityop(A,p,-1/3)
```

$$
\begin{array}{c|ccccc}
 & a_1 & a_2 & a_3 & a_4 & b \\
a_3 & -1 & -1 & 1 & 0 & -1 \\
a_2 & {}^2\!/_3 & \boxed{1} & 0 & {}^{-1}\!/_3 & {}^2\!/_3 \\
c^T & 3 & 1 & 0 & 0 & 0 \\
\end{array}
$$

$$R_1 \to R_1 + R_2$$

```
>> A = eliminationop(A,1,p,1)
```

$$
\begin{array}{c|ccccc}
 & a_1 & a_2 & a_3 & a_4 & b \\
a_3 & {}^{-1}\!/_3 & 0 & 1 & {}^{-1}\!/_3 & {}^{-1}\!/_3 \\
a_2 & {}^2\!/_3 & \boxed{1} & 0 & {}^{-1}\!/_3 & {}^2\!/_3 \\
c^T & 3 & 1 & 0 & 0 & 0 \\
\end{array}
$$

$$R_3 \to R_3 - R_2$$

```
>> A = eliminationop(A,3,p,-1)
```

$$
\begin{array}{c|ccccc}
 & a_1 & a_2 & a_3 & a_4 & b \\
a_3 & {}^{-1}\!/_3 & 0 & 1 & {}^{-1}\!/_3 & {}^{-1}\!/_3 \\
a_2 & {}^2\!/_3 & \boxed{1} & 0 & {}^{-1}\!/_3 & {}^2\!/_3 \\
c^T & {}^7\!/_3 & 0 & 0 & {}^1\!/_3 & {}^{-2}\!/_3 \\
\end{array}
$$

$$R_1 \leftrightarrow R_2$$

```
>> A = exchangeop(A,1,2)
```

$$
\begin{array}{c|ccccc}
 & a_1 & a_2 & a_3 & a_4 & b \\
a_2 & {}^2\!/_3 & 1 & 0 & {}^{-1}\!/_3 & {}^2\!/_3 \\
a_3 & {}^{-1}\!/_3 & 0 & 1 & {}^{-1}\!/_3 & {}^{-1}\!/_3 \\
c^T & {}^7\!/_3 & 0 & 0 & {}^1\!/_3 & {}^{-2}\!/_3 \\
\end{array}
$$

```
>>[A,p,q,e,B] = dual(A,B)
```

	a_1	a_2	a_3	a_4	b
a_2	$2/3$	1	0	$-1/3$	$2/3$
a_3	$-1/3$	0	1	$-1/3$	$-1/3$
c^T	$7/3$	0	0	$1/3$	$-2/3$

Since basis $B = \begin{bmatrix} a_2 & a_3 \end{bmatrix}$, therefore $x_B = \begin{bmatrix} x_2 \\ x_3 \end{bmatrix} = \begin{bmatrix} 2/3 \\ -1/3 \end{bmatrix}$. We find

pivot row p=min $\{x_B\} = \min \begin{bmatrix} 2/3 \\ -1/3 \end{bmatrix} = 2^{\text{nd}}$ row, and pivot column

q=arg min $\left\{ |\frac{7/3}{-1/3}|, \cancel{\frac{0}{0}}, \cancel{\frac{0}{1}}, |\frac{1/3}{-1/3}| \right\}$=4th column.

p = 2   q = 4   e = -1/3   B = $\begin{bmatrix} 2 & 4 \end{bmatrix}$.

	a_1	a_2	a_3	a_4	b
a_2	$2/3$	1	0	$-1/3$	$2/3$
a_4	$-1/3$	0	1	$-1/3$	$-1/3$
c^T	$7/3$	0	0	$1/3$	$-2/3$

$$R_2 \to -3R_2$$

```
>> A = identityop(A,p,-3)
```

	a_1	a_2	a_3	a_4	b
a_2	$2/3$	1	0	$-1/3$	$2/3$
a_4	1	0	-3	1	1
c^T	$7/3$	0	0	$1/3$	$-2/3$

$$R_1 \to R_1 + \tfrac{1}{3}R_2$$

```
>> A = eliminationop(A,1,p,1/3)
```

$$\begin{array}{ccccccc}
 & a_1 & a_2 & a_3 & a_4 & b \\
a_2 & 1 & 1 & -1 & 0 & 1 \\
a_4 & 1 & 0 & -3 & \boxed{1} & 1 \\
c^T & 7/3 & 0 & 0 & 1/3 & -2/3
\end{array}$$

$$R_3 \to R_3 - \tfrac{1}{3}R_2$$

```
>> A = eliminationop(A,3,p,-1/3)
```

$$\begin{array}{ccccccc}
 & a_1 & a_2 & a_3 & a_4 & b \\
a_2 & 1 & 1 & -1 & 0 & 1 \\
a_4 & 1 & 0 & -3 & 1 & 1 \\
c^T & 2 & 0 & 1 & 0 & -1
\end{array}$$

$B = \begin{bmatrix} a_2 & a_4 \end{bmatrix}, x_B = \begin{bmatrix} x_2 \\ x_4 \end{bmatrix} = \begin{bmatrix} 1 \\ 1 \end{bmatrix}$ and $b_{i0} \geq 0$. The current basic

feasible solution is optimal. Thus, $x = \begin{bmatrix} x_1 \\ x_2 \\ x_3 \\ x_4 \end{bmatrix} = \begin{bmatrix} 0 \\ 1 \\ 0 \\ 1 \end{bmatrix}$ and the opti-

mal value is 1.

Example 7.6. Use the dual simplex method to solve:

$$\begin{array}{ll}
\text{minimize} & 3x_1 + 4x_2 + 5x_3 \\
\text{subject to} & x_1 + 2x_2 + 3x_3 \geq 5, \\
 & 2x_1 + 2x_2 + x_3 \geq 6, \\
 & x_1, \quad x_2, \quad x_3 \geq 0.
\end{array}$$

In standard form:

$$\begin{array}{ll}
\text{minimize} & 3x_1 + 4x_2 + 5x_3 \\
\text{subject to} & -x_1 - 2x_2 - 3x_3 \leq -5, \\
 & -2x_1 - 2x_2 - x_3 \leq -6; \\
 & x_1, \quad x_2, \quad x_3 \geq 0.
\end{array}$$

That is,

$$\begin{aligned}
\text{minimize} \quad & 3x_1 + 4x_2 + 5x_3 \\
\text{subject to} - \quad & x_1 - 2x_2 - 3x_3 + x_4 \qquad\;\; = -5, \\
& -2x_1 - 2x_2 - \;\; x_3 \qquad\; +x_5 = -6, \\
& x_1, \quad x_2, \quad x_3, \quad x_4, \quad x_5 \geq \;\; 0.
\end{aligned}$$

The augmented matrix is

	a_1	a_2	a_3	a_4	a_5	b
a_4	-1	-2	-3	1	0	-5
a_5	-2	-2	-1	0	1	-6
c^T	3	4	5	0	0	0

Since the basis matrix is $B = \begin{bmatrix} a_4 & a_5 \end{bmatrix}$ and $x_B = \begin{bmatrix} a_4 \\ a_5 \end{bmatrix}$, therefore $Bx_B = b$ implies $x_B = \begin{bmatrix} -5 \\ -6 \end{bmatrix}$. It has a basis solution, but it is not feasible.

```
>>B=[4  5]
>>[A,p,q,e,B]=dual(A,B)
```

	a_1	a_2	a_3	a_4	a_5	b
a_4	-1	-2	-3	1	0	-5
a_5	-2	-2	-1	0	1	-6
c^T	3	4	5	0	0	0

We find pivot row $p=\min\{x_B\} = \min \begin{bmatrix} -5 \\ -6 \end{bmatrix} = 2^{\text{nd}}$ row and pivot column $q=\arg\min \left\{ |\frac{3}{-2}|, |\frac{4}{-2}|, |\frac{5}{-1}|, \cancel{\frac{0}{0}}, \cancel{\frac{0}{1}} \right\} = 1^{\text{st}}$ column.

p=2 \qquad q=1 \qquad e=-2 \qquad $B = \begin{bmatrix} 4 & 1 \end{bmatrix}$.

$$
\begin{array}{c|cccccc}
 & a_1 & a_2 & a_3 & a_4 & a_5 & b \\
a_4 & -1 & -2 & -3 & 1 & 0 & -5 \\
a_1 & \boxed{-2} & -2 & -1 & 0 & 1 & -6 \\
c^T & 3 & 4 & 5 & 0 & 0 & 0 \\
\end{array}
$$

$$R_2 \to -\tfrac{1}{2}R_2$$

```
>> A = identityop(A,p,-1/2)
```

$$
\begin{array}{c|cccccc}
 & a_1 & a_2 & a_3 & a_4 & a_5 & b \\
a_4 & -1 & -2 & -3 & 1 & 0 & -5 \\
a_1 & \boxed{1} & 1 & 1/2 & 0 & -1/2 & 3 \\
c^T & 3 & 4 & 5 & 0 & 0 & 0 \\
\end{array}
$$

$$R_1 \to R_1 + R_2$$

```
>> A = eliminationop(A,1,p,1)
```

$$
\begin{array}{c|cccccc}
 & a_1 & a_2 & a_3 & a_4 & a_5 & b \\
a_4 & 0 & -1 & -5/2 & 1 & -1/2 & -2 \\
a_1 & \boxed{1} & 1 & 1/2 & 0 & -1/2 & 3 \\
c^T & 3 & 4 & 5 & 0 & 0 & 0 \\
\end{array}
$$

$$R_3 \to R_3 - 3R_2$$

```
>> A = eliminationop(A,3,p,-3)
```

$$
\begin{array}{c|cccccc}
 & a_1 & a_2 & a_3 & a_4 & a_5 & b \\
a_4 & 0 & -1 & -5/2 & 1 & -1/2 & -2 \\
a_1 & \boxed{1} & 1 & 1/2 & 0 & -1/2 & 3 \\
c^T & 0 & 1 & 7/2 & 0 & 3/2 & -9 \\
\end{array}
$$

$$R_1 \leftrightarrow R_2$$

```
>> A = exchangeop(A,1,2)
```

	a_1	a_2	a_3	a_4	a_5	b
a_1	1	1	$1/2$	0	$-1/2$	3
a_4	0	-1	$-5/2$	1	$-1/2$	-2
c^T	0	1	$7/2$	0	$3/2$	-9

We have basis $B = \begin{bmatrix} a_1 & a_4 \end{bmatrix}$. Thus, $x_B = \begin{bmatrix} 3 \\ -2 \end{bmatrix}$. This is not feasible.

```
>> [A,p,q,e,B] = dual(A,B)
```

	a_1	a_2	a_3	a_4	a_5	b
a_1	1	1	$1/2$	0	$-1/2$	3
a_4	0	-1	$-5/2$	1	$-1/2$	-2
c^T	0	1	$7/2$	0	$3/2$	-9

We find pivot row p=min$\{x_B\}$ = min $\begin{bmatrix} \cancel{3} \\ -2 \end{bmatrix}$ = 2nd row and pivot column q=arg min $\left\{ \cancel{\frac{0}{0}}, |\frac{1}{-1}|, |\frac{7/2}{-5/2}|, \cancel{\frac{0}{1}}, |\frac{3/2}{-1/2}| \right\}$ = 2nd column.

p=2   q=2   e=-1   B=$\begin{bmatrix} 1 & 2 \end{bmatrix}$.

	a_1	a_2	a_3	a_4	a_5	b
a_1	1	1	$1/2$	0	$-1/2$	3
a_2	0	-1	$-5/2$	1	$-1/2$	-2
c^T	0	1	$7/2$	0	$3/2$	-9

$$R_2 \to -R_2$$

```
>> A = identityop(A,p,-1)
```

$$
\begin{array}{ccccccc}
 & a_1 & a_2 & a_3 & a_4 & a_5 & b \\
a_1 & 1 & 1 & 1/2 & 0 & -1/2 & 3 \\
a_2 & 0 & \boxed{1} & 5/2 & -1 & 1/2 & 2 \\
c^T & 0 & 1 & 7/2 & 0 & 3/2 & -9
\end{array}
$$

$$R_3 \to R_3 - R_2$$

```
>> A = eliminationop(A,3,p,-1)
```

$$
\begin{array}{ccccccc}
 & a_1 & a_2 & a_3 & a_4 & a_5 & b \\
a_1 & 1 & 1 & 1/2 & 0 & -1/2 & 3 \\
a_2 & 0 & \boxed{1} & 5/2 & -1 & 1/2 & 2 \\
c^T & 0 & 0 & 1 & 1 & 1 & -11
\end{array}
$$

$$R_1 \to R_1 - R_2$$

```
>> A = eliminationop(A,1,p,-1)
```

$$
\begin{array}{ccccccc}
 & a_1 & a_2 & a_3 & a_4 & a_5 & b \\
a_1 & 1 & 0 & -2 & 1 & -1 & 1 \\
a_2 & 0 & 1 & 5/2 & -1 & 1/2 & 2 \\
c^T & 0 & 0 & 1 & 1 & 1 & -11
\end{array}
$$

Since $b_{i0} \geq 0$, therefore the present solution is an optimal solution. Thus, $x_1 = 1, x_2 = 2$, and the objective value is 11.

Example 7.7. Consider the linear programming problem

$$
\begin{aligned}
\text{maximize} & -4x_1 - 6x_2 - 5x_3 \\
\text{subject to} \quad & 2x_1 \qquad\quad +3x_3 \geq 3, \\
& \qquad 3x_2 + 2x_3 \geq 6, \\
& x_1, \quad x_2, \quad x_3, \geq 0.
\end{aligned}
$$

Solve the problem by the dual simplex method.
We express the problem in standard form as

$$
\begin{aligned}
\text{minimize} \quad & 4x_1 + 6x_2 + 5x_3 \\
\text{subject to} \quad -2x_1 \qquad & -3x_3 \leq -3, \\
 -3x_2 & - 2x_3 \leq -6, \\
x_1, \quad x_2, \quad & x_3 \geq 0.
\end{aligned}
$$

That is,

$$\begin{aligned}
\text{minimize} \quad & 4x_1 + 6x_2 + 5x_3 \\
\text{subject to} \ -2x_1 \quad & \quad -3x_3 + x_4 \quad = -3, \\
& -3x_2 - 2x_3 \quad + x_5 = -6, \\
x_1, \quad x_2, \quad & x_3 \ , x_4, \ x_5 \geq \quad 0.
\end{aligned}$$

The augmented matrix is

	a_1	a_2	a_3	a_4	a_5	b
a_4	-2	0	-3	1	0	-3
a_5	0	-3	-2	0	1	-6
c^T	4	6	5	0	0	0

Since basis matrix $B = \begin{bmatrix} a_4 & a_5 \end{bmatrix}$ and $x_B = \begin{bmatrix} x_4 \\ x_5 \end{bmatrix}$, therefore, the $Bx_B = b$ implies $x_B = \begin{bmatrix} -3 \\ -6 \end{bmatrix}$. It has a basis solution, but it is not feasible.

```
>> B =[4  5]
>> [A,p,q,e,B] = dual(A,B)
```

	a_1	a_2	a_3	a_4	a_5	b
a_4	-2	0	-3	1	0	-3
a_5	0	-3	-2	0	1	-6
c^T	4	6	5	0	0	0

We find pivot row p=min $\{x_B\}$ = min $\begin{bmatrix} -3 \\ -6 \end{bmatrix}$ = 2nd row and the

pivot column q=arg min $\left\{ \cancel{\tfrac{4}{0}}, \left| \tfrac{6}{-3} \right|, \left| \tfrac{5}{-2} \right|, \cancel{\tfrac{0}{0}}, \cancel{\tfrac{0}{1}} \right\}$= 2nd column.

```
p = 2    q = 2    e = -3    B =[4  2] .
```

$$
\begin{array}{c|cccccc}
 & a_1 & a_2 & a_3 & a_4 & a_5 & b \\
a_4 & -2 & 0 & -3 & 1 & 0 & -3 \\
a_2 & 0 & \boxed{-3} & -2 & 0 & 1 & -6 \\
c^T & 4 & 6 & 5 & 0 & 0 & 0
\end{array}
$$

$$R_2 \to -\tfrac{1}{3} R_2$$

```
>> A = identityop(A,p,-1/3)
```

$$
\begin{array}{c|cccccc}
 & a_1 & a_2 & a_3 & a_4 & a_5 & b \\
a_4 & -2 & 0 & -3 & 1 & 0 & -3 \\
a_2 & 0 & \boxed{1} & 2/3 & 0 & -1/3 & 2 \\
c^T & 4 & 6 & 5 & 0 & 0 & 0
\end{array}
$$

$$R_3 \to R_3 - 6R_2$$

```
>> A = eliminationop(A,3,p,-6)
```

$$
\begin{array}{c|cccccc}
 & a_1 & a_2 & a_3 & a_4 & a_5 & b \\
a_4 & -2 & 0 & -3 & 1 & 0 & -3 \\
a_2 & 0 & 1 & 2/3 & 0 & -1/3 & 2 \\
c^T & 4 & 0 & 1 & 0 & 2 & -12
\end{array}
$$

We have available basis $B = \begin{bmatrix} a_4 & a_2 \end{bmatrix}$. Thus, $x_B = \begin{bmatrix} -3 \\ 2 \end{bmatrix}$. It is basic, but it does not give a feasible solution.

```
>> [A,p,q,e,B] = dual(A,B)
```

$$
\begin{array}{c|cccccc}
 & a_1 & a_2 & a_3 & a_4 & a_5 & b \\
a_4 & -2 & 0 & -3 & 1 & 0 & -3 \\
a_2 & 0 & 1 & 2/3 & 0 & -1/3 & 2 \\
c^T & 4 & 0 & 1 & 0 & 2 & -12
\end{array}
$$

We find pivot row p=min $\{x_B\}$ = min $\begin{bmatrix} -3 \\ 2 \end{bmatrix}$ =1st row and pivot column q=arg min $\left\{ |\tfrac{4}{-2}|, \cancel{\tfrac{0}{0}}, |\tfrac{1}{-3}|, \cancel{\tfrac{0}{1}}, \cancel{\tfrac{2}{0}} \right\}$ = 3rd column.

p = 1   q = 3   e = -3   B $=\begin{bmatrix} 3 & 2 \end{bmatrix}$.

	a_1	a_2	a_3	a_4	a_5	b
a_3	-2	0	$\boxed{-3}$	1	0	-3
a_2	0	1	$2/3$	0	$-1/3$	2
c^T	4	0	1	0	2	-12

$$R_1 \to \tfrac{-1}{3}R_1$$

```
>> A = identityop(A,p,-1/3)
```

	a_1	a_2	a_3	a_4	a_5	b
a_3	$2/3$	0	$\boxed{1}$	$-1/3$	0	1
a_2	0	1	$2/3$	0	$-1/3$	2
c^T	4	0	1	0	2	-12

$$R_2 \to R_2 - \tfrac{2}{3}R_1$$

```
>> A = eliminationop(A,2,p,-2/3)
```

	a_1	a_2	a_3	a_4	a_5	b
a_3	$2/3$	0	$\boxed{1}$	$-1/3$	0	1
a_2	$-4/9$	1	0	$2/9$	$-1/3$	$4/3$
c^T	4	0	1	0	2	-12

$$R_3 \to R_3 - R_1$$

```
>> A = eliminationop(A,3,p,-1)
```

	a_1	a_2	a_3	a_4	a_5	b
a_3	$2/3$	0	1	$-1/3$	0	1
a_2	$-4/9$	1	0	$2/9$	$-1/3$	$4/3$
c^T	$10/3$	0	0	$1/3$	2	-13

Since all $b_{i0} \geq 0$, therefore, the present solution is optimal. We have

basis $B = \begin{bmatrix} a_3 & a_2 \end{bmatrix}$, $x_B = \begin{bmatrix} x_3 \\ x_2 \end{bmatrix} = \begin{bmatrix} 1 \\ 4/3 \end{bmatrix}$ and $x = \begin{bmatrix} x_1 \\ x_2 \\ x_3 \end{bmatrix} = \begin{bmatrix} 0 \\ 4/3 \\ 1 \end{bmatrix}$ is

the optimal solution. The minimum objective value is 13.

Example 7.8. Consider the following linear programming problem:

$$
\begin{aligned}
\text{minimize} \quad & 160x_1 + 400x_2 + 300x_3 \\
\text{subject to} \quad & 3x_1 + 6x_2 + 6x_3 \geq 36, \\
& 4x_1 + 6x_2 + 3x_3 \geq 20, \\
& 2x_1 + 8x_2 + 4x_3 \geq 30, \\
& x_1, \quad x_2, \quad x_3 \geq 0.
\end{aligned}
$$

Solve using the dual simplex method.
Standard form of the linear programming problem:

$$
\begin{aligned}
\text{minimize} \quad & 160x_1 + 400x_2 + 300x_3 \\
\text{subject to} \quad & - 3x_1 - 6x_2 - 6x_3 \leq -36, \\
& - 4x_1 - 6x_2 - 3x_3 \leq -20, \\
& - 2x_1 - 8x_2 - 4x_3 \leq -30, \\
& x_1, \quad x_2, \quad x_3 \geq 0.
\end{aligned}
$$

That is,

$$
\begin{aligned}
\text{minimize} \quad & 160x_1 + 400x_2 + 300x_3 \\
\text{subject to} \quad & - 3x_1 - 6x_2 - 6x_3 + x_4 & = -36, \\
& - 4x_1 - 6x_2 - 3x_3 & +x_5 & = -20, \\
& - 2x_1 - 8x_2 - 4x_3 & +x_6 & = -30, \\
& x_1, \quad x_2, \quad x_3 \quad x_4, \quad x_5, \quad x_6 \geq 0.
\end{aligned}
$$

The augmented matrix is

	a_1	a_2	a_3	a_4	a_5	a_6	b
a_4	-3	-6	-6	1	0	0	-36
a_5	-4	-6	-3	0	1	0	-20
a_6	-2	-8	-4	0	0	1	-30
c^T	160	400	300	0	0	0	0

Note that the basis matrix $B = \begin{bmatrix} a_4 & a_5 & a_6 \end{bmatrix}$, $x_B = \begin{bmatrix} x_4 \\ x_5 \\ x_6 \end{bmatrix}$, and

$Bx_B = b$ implies $x_B = \begin{bmatrix} -36 \\ -20 \\ -30 \end{bmatrix}$.

```
>> B = [4 5 6]
>> [A,p,q,e,B] = dual(A,B)
```

	a_1	a_2	a_3	a_4	a_5	a_6	b
a_3	-3	-6	-6	1	0	0	-36
a_5	-4	-6	-3	0	1	0	-20
a_6	-2	-8	-4	0	0	1	-30
c^T	160	400	300	0	0	0	0

```
p = 1    q = 3    e = -6    B = [3 5 6].
```

```
>> A = simplex(A,p,q)
```

	a_1	a_2	a_3	a_4	a_5	a_6	b
a_3	$1/2$	1	1	$-1/6$	0	0	6
a_5	$-5/2$	-3	0	$-1/2$	1	0	-2
a_6	0	-4	0	$-2/3$	0	1	-6
c^T	10	100	0	50	0	0	-18000

```
>> [A,p,q,e,B] = dual(A,B)
```

	a_1	a_2	a_3	a_4	a_5	a_6	b
a_3	$1/2$	1	1	$-1/6$	0	0	6
a_5	$-5/2$	-3	0	$-1/2$	1	0	-2
a_6	0	-4	0	$-2/3$	0	1	-6
c^T	10	100	0	50	0	0	-18000

```
p=3    q=2    e=-4    B=[3 5 2].
```

	a_1	a_2	a_3	a_4	a_5	a_6	b
a_3	$1/2$	0	1	$-1/3$	0	$1/4$	$9/2$
a_5	$-5/2$	0	0	0	1	$-3/4$	$5/2$
a_2	0	1	0	$1/6$	0	$-1/4$	$3/2$
c^T	10	0	0	$100/3$	0	25	-1950

All $b_{i0} \geq 0$, which yields an optimal solution as $x_B = \begin{bmatrix} x_1 \\ x_2 \\ x_3 \end{bmatrix} =$

$\begin{bmatrix} 0 \\ 3/2 \\ 9/2 \end{bmatrix}$ and value of the objective function is 1950.

Example 7.9. If P and D are a primal-dual pair of linear programming, then which of the following statements is FALSE?

(a) If P has an optimal solution, then D also has an optimal solution.

(b) The dual of the dual problem is a primal problem.

(c) If P has an unbounded solution, then D has no feasible solution.

(d) If P has no feasible solution, then D has a feasible solution.

Option (d) is false. If P has no feasible solution, then D has a feasible solution.

Example 7.10. For a linear programming primal maximization problem P with dual Q, which of the following statements is TRUE?

(a) The optimal values of P and Q exist and are the same.

(b) Both optimal values exist, and the optimal value of P is less than the optimal value of Q.

(c) P will have an optimal solution, if and only if Q also has an optimal solution.

(d) Both P and Q cannot be infeasible.

For a linear programming primal maximization problem P with dual Q, P will have an optimal solution, if and only if Q also has an optimal solution. Therefore, option (c) is true.

Example 7.11. Suppose that the linear programming problem

$$
\begin{array}{rl}
\text{maximize} & c^T x \\
\text{subject to} & Ax \le b \\
& x \ge 0
\end{array}
$$

admits a feasible solution and the dual

$$
\begin{array}{rl}
\text{maximize} & b^T y \\
\text{subject to} & A^T y \ge c \\
& y \ge 0
\end{array}
$$

admits a feasible solution y_0. Then,

(a) the dual admits an optimal solution.

(b) any feasible solution of the primal and of the dual satisfies.

(c) the dual problem is unbounded.

(d) the primal problem admits an optimal solution.

Options (a,d) are true. That is, the primal and dual both admit an optimal solution.

7.4   Exercises

Exercise 7.1. Determine the dual of each of the following linear programming problems.

(a)

$$
\begin{array}{ll}
\text{maximize} & 10x_1 + 30x_2 \\
\text{subject to} & 5x_1 - 4x_2 \leq 100, \\
& x_1 + 12x_2 \leq 90, \\
& x_2 \leq 400, \\
& x_1, x_2 \geq 0.
\end{array}
$$

(b)

$$
\begin{array}{ll}
\text{minimize} & 3x_1 - 4x_2 \\
\text{subject to} & 6x_1 + 11x_2 \geq -30, \\
& 2x_1 - 7x_2 \leq 50, \\
& x_2 \leq 80, \\
& x_1, x_2 \geq 0.
\end{array}
$$

(c)

$$
\begin{array}{ll}
\text{maximize} & - x_1 + 2x_2 \\
\text{subject to} & 5x_1 + x_2 \leq 60, \\
& 3x_1 - 8x_2 \geq 10, \\
& x_1 + 7x_2 = 20, \\
& x_1, x_2 \geq 0.
\end{array}
$$

Exercise 7.2. Consider the linear programming problem.

$$
\begin{array}{ll}
\text{minimize} & - 4x_1 - 3x_2 - 2x_3 \\
\text{subject to} & 2x_1 + 3x_2 + 2x_3 \leq 6, \\
& - x_1 + x_2 + x_3 \leq 5, \\
& x_1, x_2, x_3 \geq 0.
\end{array}
$$

(a) Write down the dual of this linear programming problem.

(b) Solve the primal problem by the simplex method.

Exercise 7.3. Solve the linear programming problem using the dual simplex method.

$$\text{minimize} \quad 2x_1 + x_2$$

$$
\begin{aligned}
\text{subject to} \quad -3x_1 - x_2 + x_3 \qquad\qquad &= -3, \\
-4x_1 - 3x_2 \qquad + x_4 \qquad &= -6, \\
-x_1 - 2x_2 \qquad\qquad + x_5 &= -2, \\
x_1, \quad x_2, \quad x_3, \quad x_4, \quad x_5 &\geq 0.
\end{aligned}
$$

Exercise 7.4. Solve the linear programming problem using the dual simplex method.

$$\text{minimize} \qquad x_1 + 45x_2 + 3x_3$$

$$
\begin{aligned}
\text{subject to} \qquad x_1 + 5x_2 - x_3 &\geq 4, \\
x_1 + x_2 + 2x_3 &\geq 2, \\
-x_1 + 3x_2 + 3x_3 &\geq 5, \\
-3x_1 + 8x_2 - 5x_3 &\geq 3, \\
x_1, \quad x_2, \quad x_3 &\geq 0.
\end{aligned}
$$

Exercise 7.5. Solve the linear programming problem using the dual simplex method.

$$\text{minimize} \quad 10x_1 + 2x_2 + 4x_3 + 8x_4 + x_5$$

$$
\begin{aligned}
\text{subject to} \quad x_1 + 4x_2 - x_3 \qquad\qquad &\geq 16, \\
2x_1 + x_2 + x_3 \qquad\qquad &\geq 4, \\
3x_1 \qquad\qquad + x_4 - x_5 &\geq 8, \\
x_1 \qquad\qquad + x_4 - x_5 &\geq 20, \\
x_1, \quad x_2, \quad x_3, \quad x_4, \quad x_5 &\geq 0.
\end{aligned}
$$

Exercise 7.6. Solve the linear programming problem.

$$\text{minimize} \quad 2x_1 + 3x_2$$

$$
\begin{aligned}
\text{subject to} \quad 4x_1 - 3x_2 &\geq 5, \\
x_1 + 2x_2 &\geq 4, \\
x_1, \quad x_2 &\geq 0.
\end{aligned}
$$

Exercise 7.7. Consider the following linear programming problem.

$$\text{minimize} \quad 3x_1 + 4x_2 + 5x_3$$

$$
\begin{aligned}
\text{subject to} \quad x_1 + 3x_2 + x_3 &\geq 2, \\
2x_1 - x_2 + 3x_3 &\geq 3, \\
x_1, \quad x_2, \quad x_3 &\geq 0.
\end{aligned}
$$

Solve using

(a) the dual simplex method.

(b) the simplex method on the dual of the problem.

Exercise 7.8. Solve the linear programming problem using the dual simplex method.

$$
\begin{aligned}
\text{minimize} \quad & 17x_1 + 7x_2 + 17x_3 \\
\text{subject to} \quad & x_1 + x_2 \geq 8, \\
& x_1 + 4x_2 + 2x_3 \geq 14, \\
& 3x_2 + 4x_3 \geq 9, \\
& x_1, \quad x_2, \quad x_3 \geq 0.
\end{aligned}
$$

Exercise 7.9. Solve the linear programming problem using the dual simplex method.

$$
\begin{aligned}
\text{minimize} \quad & 30x_1 + 50x_2 + 26x_3 \\
\text{subject to} \quad & 2x_1 + 0.5x_2 + x_3 \geq 25, \\
& x_1 + 3x_2 + 2x_3 \geq 40, \\
& 2x_1 + x_2 + x_3 \geq 30, \\
& x_1, \quad x_2, \quad x_3 \geq 0.
\end{aligned}
$$

Exercise 7.10. Solve the linear programming problem using the dual simplex method.

$$
\begin{aligned}
\text{minimize} \quad & 10x_1 + 4x_2 \\
\text{subject to} \quad & 3x_1 + 2x_2 \geq 60, \\
& 7x_1 + 2x_2 \geq 84, \\
& 3x_1 + 6x_2 \geq 72, \\
& x_1, \quad x_2 \geq 0.
\end{aligned}
$$

Exercise 7.11. Solve the following problem by the dual simplex method.

$$
\begin{aligned}
\text{minimize} \quad & x_1 + 3x_2 + 4x_3 + x_4 + 2x_5 \\
\text{subject to} \quad & 2x_1 + 5x_2 + 3x_3 - 2x_4 + 6x_5 \geq 10, \\
& - x_1 - 2x_2 - 4x_3 + x_4 + 2x_5 \geq 12, \\
& x_1, \quad x_2, \quad x_3, \quad x_4, \quad x_5 \geq 0.
\end{aligned}
$$

Exercise 7.12. Solve the linear programming problem using the dual simplex method.

$$\begin{array}{ll} \text{minimize} & 8x_1 + 8x_2 + 16x_3 + 7x_4 \\ \text{subject to} & x_1 + 3x_2 + 3x_3 + 2x_4 \geq 32, \\ & 2x_1 + 2x_2 + 8x_3 + 3x_4 \geq 28, \\ & 7x_1 + 4x_2 + 6x_3 + 5x_4 \geq 35, \\ & x_1, \quad x_2, \quad x_3, \quad x_4 \geq 0. \end{array}$$

Exercise 7.13. Solve the following problems using the dual simplex method.

$$\begin{array}{ll} \text{minimize} & x_1 + x_2 \\ \text{subject to} & 2x_1 + x_2 \geq 1, \\ & x_1 + 2x_2 \geq 1, \\ & x_1, \quad x_2 \geq 0, \end{array}$$

and hence

$$\begin{array}{ll} \text{minimize} & x_1 + x_2 + x_3 + \frac{1}{3}x_4 \\ \text{subject to} & 2x_1 + x_2 + 2x_3 \qquad \geq 1, \\ & x_1 + 2x_2 \qquad +2x_4 \geq 1, \\ & 2x_1 + 2x_2 \qquad + x_4 \geq 1, \\ & x_1, \quad x_2, \quad x_3, \quad x_4 \geq 0. \end{array}$$

Chapter 8

The Transportation Problem

8.1 Introduction

Transportation means the transfer of materials from different sources to different destinations. Suppose that a firm has production units at $O_1, O_2, \ldots, O_m$ places. The demand for produced goods is at n different centers $D_1, D_2 \ldots, D_n$. The problem of the firm is to transport goods from m different production units to n different demand centers with minimum cost. Consider the cost of shipping from production unit O_i to the demand center D_j is c_{ij}, and x_{ij} unit is shipped from O_i to D_j, then the cost is $c_{ij}x_{ij}$. Therefore, the total shipping cost is

$$z = \sum_{i=1}^{m} \sum_{j=1}^{n} c_{ij}x_{ij}. \tag{8.1}$$

Note that z is a linear. From (8.1), the matrix $(c_{ij})_{m \times n}$ is called the "unit cost matrix". The goods are transferred from the source i to the demand center j. We wish to find $x_{ij} \geq 0$ which satisfy the $m + n$ constraints.

Then, we have

$$\sum_{i=1}^{m} a_i = a, \tag{8.2}$$

$$\sum_{j=1}^{n} b_j = b, \tag{8.3}$$

where a and b are total supply and demand. The problem of transportation is to find x_{ij} so that the cost of transportation z is minimum. If the amount of goods available at the ith source is transferred to j^{th} destination, then

$$\sum_{j=1}^{n} x_{ij} = a_i, \tag{8.4}$$

$$\sum_{i=1}^{m} x_{ij} = b_j. \tag{8.5}$$

Using (8.1) and (8.4)–(8.5), the transportation problem can be formulated as a linear programming problem

$$
\begin{aligned}
\text{minimize} \quad & z = \sum_{i=1}^{m}\sum_{j=1}^{n} c_{ij} x_{ij} \\
\text{subject to} \quad & \sum_{j=1}^{n} x_{ij} = a_i, \qquad i = 1, 2, \ldots, m, \\
& \sum_{i=1}^{m} x_{ij} = b_j, \qquad j = 1, 2, \ldots, n, \\
& x_{ij} \geq 0 \quad \forall i, j.
\end{aligned}
\tag{8.6}
$$

Equivalently,

$$
\begin{aligned}
\text{minimize} \quad & c_{11} x_{11} + c_{12} x_{12} + \cdots + c_{mn} x_{mn} \\
\text{subject to} \quad & x_{11} + \quad x_{12} + \cdots + \quad x_{1n} = a_1, \\
& x_{21} + \quad x_{22} + \cdots + \quad x_{2n} = a_2, \\
& x_{31} + \quad x_{32} + \cdots + \quad x_{3n} = a_3, \\
& \qquad \vdots \\
& x_{m1} + \quad x_{m2} + \cdots + \quad x_{mn} = a_m, \\
& x_{11} + \quad x_{21} + \cdots + \quad x_{m1} = b_1, \\
& x_{12} + \quad x_{22} + \cdots + \quad x_{m2} = b_2, \\
& x_{13} + \quad x_{23} + \cdots + \quad x_{m3} = b_3, \\
& \qquad \vdots \\
& x_{1n} + \quad x_{2n} + \cdots + \quad x_{mn} = b_n, \\
& x_{ij} \geq 0, i = 1, 2, \ldots, m, j = 1, 2, \ldots, n.
\end{aligned}
$$

That is,

$$
\begin{aligned}
\text{minimize} \quad & c^T x \\
\text{subject to} \quad & Ax = b, \\
& x \geq 0,
\end{aligned}
$$

where

$$c^T = \begin{bmatrix} c_{11} & c_{12} & \dots & c_{mn} \end{bmatrix},$$
$$x^T = \begin{bmatrix} x_{11} & x_{12} & \dots & x_{1n} & x_{21} & \dots & x_{2n} & \dots & x_{mn} \end{bmatrix},$$
$$b^T = \begin{bmatrix} a_1 & a_2 & \dots & a_m & b_1 & b_2 & \dots & b_n \end{bmatrix},$$
$$A = \begin{bmatrix} a_{11} & a_{12} & \dots & a_{mn} \end{bmatrix},$$
$$a_{11}^T = \begin{bmatrix} 1 & 0 & \dots & 0 & 1 & 0 & \dots & 0 \end{bmatrix},$$
$$a_{mn}^T = \begin{bmatrix} 0 & \dots & 1 & 0 & \dots & 1 \end{bmatrix}.$$

8.2 Balanced Transportation Problem

If the total quantity required at destinations is precisely the same as the amount available at the origins, then the problem is said to be a balanced transportation problem. Therefore, using (8.2) and (8.3) to get

$$\sum_{i=1}^{m} a_i = \sum_{j=1}^{n} b_j. \tag{8.7}$$

The transportation problem is a special type of linear programming problem. Thus, the definition of basic feasible solution of the transportation problem is the same as the definition of the linear programming problem.

Frrom (8.6), it follows that x_{ij} are known as decision variables. They are mn in total. But, the number of basic variables is much less than mn in a transportation problem.

Theorem 8.1. *In a balanced transportation problem, there are at most $m + n - 1$ basic variables.*

Proof. Consider a balanced transportation problem

$$
\begin{aligned}
\text{minimize} \quad & z = \sum_{i=1}^{m} \sum_{j=1}^{n} c_{ij} x_{ij} \\
\text{subject to} \quad & \sum_{j=1}^{n} x_{ij} = a_i, \qquad i = 1, 2, \ldots, m, \\
& \sum_{i=1}^{m} x_{ij} = b_j, \qquad j = 1, 2, \ldots, n. \\
& x_{ij} \geq 0 \ \forall \ i, j.
\end{aligned} \tag{8.8}
$$

However,

$$
\sum_{i=1}^{m} a_i = \sum_{j=1}^{n} b_j. \tag{8.9}
$$

Note that the transportation problem has $m+n$ linear constraints with mn variables. In order to show that there are $m+n-1$ basic variables, we must show that out of $m+n$ linear constraints, only $m+n-1$ are linear independent. For that, it is sufficient to show that any one of $m+n$ linear constraints can be written as a linear combination of the other linear constraints.

Summing the m constraints of (8.4) to get

$$
\sum_{i=1}^{m} \sum_{j=1}^{n} x_{ij} = \sum_{i=1}^{m} a_i = \sum_{j=1}^{n} b_j. \tag{8.10}
$$

The transportation problem is balanced.
Summing the first $n-1$ of (8.5) to get

$$
\sum_{j=1}^{n-1} \sum_{i=1}^{m} x_{ij} = \sum_{j=1}^{n-1} b_j. \tag{8.11}
$$

Subtracting (8.11) from (8.10) to obtain

$$
\sum_{i=1}^{m} \sum_{j=1}^{n} x_{ij} - \sum_{j=1}^{n-1} \sum_{i=1}^{m} x_{ij} = \sum_{i=1}^{m} a_i - \sum_{j=1}^{n-1} b_j
$$

$$
= \sum_{j=1}^{n} b_j - \sum_{j=1}^{n-1} b_j
$$

$$
= b_n.
$$

That is,

$$\sum_{i=1}^{m}\sum_{j=1}^{n} x_{ij} - \sum_{j=1}^{n-1}\sum_{i=1}^{m} x_{ij} = b_n$$

$$\sum_{i=1}^{m}\left[\sum_{j=1}^{n} x_{ij} - \sum_{j=1}^{n-1} x_{ij}\right] = b_n$$

$$\sum_{i=1}^{m} x_{in} = b_n.$$

It is the 2nd linear equation of (8.8). □

Note: If a feasible solution involves exactly $m + n - 1$ independent positive allocations, then it is a nondegenerate basic feasible solution, otherwise it is said to be a degenerate basic feasible solution.

Theorem 8.2. *There exists a feasible solution to the transportation problem if and only if*

$$\sum_{i=1}^{m} a_i = \sum_{j=1}^{n} b_j.$$

Proof. Recall the transportation problem (8.6)

$$\text{minimize} \quad z = \sum_{i=1}^{m}\sum_{j=1}^{n} c_{ij} x_{ij}$$

$$\text{subject to} \quad \sum_{j=1}^{n} x_{ij} = a_i, \qquad i = 1, 2, \ldots, m,$$

$$\sum_{i=1}^{m} x_{ij} = b_j, \qquad j = 1, 2, \ldots, n,$$

$$x_{ij} \geq 0 \ \forall \ i, j.$$

Suppose that a feasible solution to the above transportation problem exists, then we get

$$\sum_{i=1}^{m} a_i = \sum_{i=1}^{m} \sum_{j=1}^{n} x_{ij}$$

$$= \sum_{j=1}^{n} \left(\sum_{i=1}^{m} x_{ij} \right)$$

$$= \sum_{j=1}^{n} b_j.$$

That is,

$$\sum_{i=1}^{m} a_i = \sum_{j=1}^{n} b_j.$$

Conversely, assume that

$$\sum_{i=1}^{m} a_i = \sum_{j=1}^{n} b_j = \lambda \ .$$

We know that $a_i \geq 0$, where $i = 1, 2, \ldots, m$, that is, available goods at 'm' source centers and $b_j \geq 0$, where $j = 1, 2, \ldots, n$, that is, demand at 'n' destination centers.
Note that all a_i and b_j cannot be zero. Therefore, $\lambda > 0$. Let

$$\frac{a_i b_j}{\lambda} = y_{ij}.$$

Since $a_i \geq 0$, $b_j \geq 0$ and $\lambda > 0$, we have $y_{ij} \geq 0$.
Indeed,

$$\sum_{j=1}^{n} y_{ij} = \sum_{j=1}^{n} \frac{a_i b_j}{\lambda}$$

$$= \frac{a_i}{\lambda} \sum_{j=1}^{n} b_j$$

$$= \frac{a_i}{\lambda} \lambda.$$

That is,

$$\sum_{j=1}^{n} y_{ij} = a_i.$$

Similarly, we can also have

$$\sum_{i=1}^{m} y_{ij} = \sum_{i=1}^{m} \frac{a_i b_j}{\lambda}$$

$$= \frac{b_j}{\lambda} \sum_{i=1}^{m} a_i$$

$$= \frac{b_j}{\lambda} \lambda.$$

That is,

$$\sum_{i=1}^{m} y_{ij} = b_j.$$

Thus, y_{ij}, where $i = 1, 2, \ldots, m$; $j = 1, 2, \ldots, n$ is a feasible solution to the transportation problem. □

Remarks : An unbalanced transportation problem can be made a balanced transportation problem. If

$$\sum_{i=1}^{m} a_i > \sum_{j=1}^{n} b_j, \tag{8.12}$$

then we can create an additional demand center, called a fictitious or dummy demand center, having an additional demand of $\sum_{i=1}^{m} a_i - \sum_{j=1}^{n} b_j$ and the transportation cost of this demand from source units is zero. This reduces an unbalanced transportation problem to a balanced transportation problem.

If

$$\sum_{i=1}^{m} a_i < \sum_{j=1}^{n} b_j, \tag{8.13}$$

then we consider a fictitious source center with goods $\sum_{j=1}^{n} b_j - \sum_{i=1}^{m} a_i$ at the fictitious origin and the transportation of $\sum_{j=1}^{n} b_j - \sum_{i=1}^{m} a_i$ from fictitious origin will be taken as a short supply.

We now study the following three methods to find the initial basic feasible solution to a transportation problem:

1. Northwest Corner Method

2. Least Cost Method

3. Vogel's Approximation Method

8.3   Northwest Corner Method

The major advantage of this method is that it is very simple and easy to apply. The algorithm to find a starting basic feasible solution is given below:

Algorithm

1. Allocate min $\{a_i, b_j\}$ to the northwest corner of the cost matrix, where a_i is available supply at the ith source and b_j is demand at the jth destination.

2. The row or column which is satisfied is ignored for further consideration. Adjust the supply and demand by subtracting the allocated amount.

3. Perform the following operations:

    (a) If the supply for the first row is satisfied, then move down in the 1st column and go to step 1.

    (b) If the demand for the 1st column is satisfied then move horizontally to the next cell in the same row and go to step 1.

4. If both row and column tend to zero simultaneously, then ignore both the row and column with respect to the allocated cell.

5. Repeat steps 3 to 4 until all the allocations are made, i.e., until the supply meets demand.

Example 8.1. Find the initial basic feasible solution of the following transportation problem.

	D_1	D_2	D_3	D_4	
O_1	5	2	4	3	30
O_2	6	4	9	5	40
O_3	2	3	8	1	55
	15	20	40	50	

Supply   : $a_1 = 30, a_2 = 40, a_3 = 55$.
Demand   : $b_1 = 15, b_2 = 20, b_3 = 40, b_4 = 50$.
Matrix cost: $c_{11} = 5, c_{12} = 2, c_{13} = 4, c_{14} = 3, c_{21} = 6, c_{22} = 4, c_{23} = 9, c_{24} = 5, c_{31} = 2, c_{32} = 3, c_{33} = 8, c_{34} = 1$.

We have

$$\sum_{i=1}^{3} a_i = a_1 + a_2 + a_3 = 125,$$

$$\sum_{j=1}^{4} b_j = b_1 + b_2 + b_3 + b_4 = 125.$$

Since

$$\sum_{i=1}^{3} a_i = \sum_{j=1}^{4} b_j,$$

therefore the problem is the balanced transportation problem. The northwest corner is x_{11}.
$x_{11} = \min \{a_1, b_1\} = \{30, 15\} = 15$. Subtract 15 from a_1 and b_1. Therefore, next transportation matrix is

	D_1	D_2	D_3	D_4	
O_1	5	2	4	3	15
	15				
O_2	6	4	9	5	40
O_3	2	3	8	1	55
	0	20	40	50	

Leave the 1st column because it is satisfied. x_{12} is the northwest corner. Thus, $x_{12} = \min\{a_1, b_2\} = \min\{15, 20\} = 15$. Subtract 15 from a_1 and b_2.

	D_1	D_2	D_3	D_4	
O_1	5	2	4	3	0
	15	15			
O_2	6	4	9	5	40
O_3	2	3	8	1	55
	0	5	40	50	

We get $x_{11} = 15$ and $x_{12} = 15$. Thus, for the next allocation, ignore 1st row and 1st column. That is, $x_{22} = \min\{a_2, b_2\} = \{40, 5\} = 5$.

	D_1	D_2	D_3	D_4	
O_1	5	2	4	3	0
	15	15			
O_2	6	4	9	5	35
		5			
O_3	2	3	8	1	55
	0	0	40	50	

For next the allocation, the 2nd column is ignored. Therefore, northwest is $x_{23} = \min\{a_2, b_3\} = \{35, 40\} = 35$.

	D_1	D_2	D_3	D_4	
O_1	5 15	2 15	4	3	0
O_2	6	4 5	9 35	5	0
O_3	2	3	8	1	55
	0	0	5	50	

We get $x_{23} = 35$. Ignore the 2nd row. The northwest cost is $x_{33} =$ min $\{a_3, b_3\} =$ min $\{55, 5\} = 5$.

	D_1	D_2	D_3	D_4	
O_1	5 15	2 15	4	3	0
O_2	6	4 5	9 35	5	0
O_3	2	3	8 5	1	50
	0	0	0	50	

Thus, $x_{33} = 5$. The last northwest corner is $x_{34} =$ min $\{a_3, b_4\} =$ min $\{50, 50\} = 50$.

	D_1	D_2	D_3	D_4	
O_1	5 15	2 15	4	3	0
O_2	6	4 5	9 35	5	0
O_3	2	3	8 5	1 50	0
	0	0	0	0	

Thus, the initial basic feasible solution of the transportation problem is given by

$$x_{11} = 15, x_{12} = 15, x_{22} = 5, x_{23} = 35, x_{33} = 5, x_{34} = 50.$$

The basic feasible solution is nondegenerate as there are $m + n - 1 = 3 + 4 - 1 = 6$ allocated cells. The corresponding minimum transportation cost is given by

$$z = 5 \times 15 + 2 \times 15 + 4 \times 5 + 9 \times 35 + 8 \times 5 + 1 \times 50$$
$$= 75 + 30 + 20 + 315 + 40 + 50$$
$$= 530 \text{ units.}$$

Example 8.2. Solve the following transportation problem using the northwest corner method.

	D_1	D_2	D_3	D_4	
O_1	19	30	50	10	7
O_2	70	30	40	60	9
O_3	40	8	70	20	18
	5	8	7	14	

Supply   : $a_1 = 7, a_2 = 9, a_3 = 18$.
Demand   : $b_1 = 5, b_2 = 8, b_3 = 7, b_4 = 14$.
Matrix cost : $c_{11} = 19, c_{12} = 30, c_{13} = 50, c_{14} = 10, c_{21} = 70,$
$c_{22} = 30, c_{23} = 40, c_{24} = 60, c_{31} = 40, c_{32} = 8,$
$c_{33} = 70, c_{34} = 20.$
We have

$$\sum_{i=1}^{3} a_i = a_1 + a_2 + a_3 = 34,$$

$$\sum_{j=1}^{4} b_j = b_1 + b_2 + b_3 + b_4 = 34.$$

Since

$$\sum_{i=1}^{3} a_i = \sum_{j=1}^{4} b_j,$$

therefore the problem is the balanced transportation problem. The northwest corner is x_{11}. We get $x_{11} = \min\{a_1, b_1\} = \min\{7, 5\} = 5$.

	D_1	D_2	D_3	D_4	
O_1	19 5	30	50	10	2
O_2	70	30	40	60	9
O_3	40	8	70	20	18
	0	8	7	14	

We move to x_{12}. Therefore, we find $\min\{2,8\}=2$.

	D_1	D_2	D_3	D_4	
O_1	19 5	30 2	50	10	0
O_2	70	30	40	60	9
O_3	40	8	70	20	18
	0	6	7	14	

The 1st row has supply zero. We move to cost cell x_{22} and find $\min\{9,6\}=6$.

	D_1	D_2	D_3	D_4	
O_1	19 5	30 2	50	10	0
O_2	70	30 6	40	60	3
O_3	40	8	70	20	18
	0	0	7	14	

The 2nd column has demand zero. We move to cost cell $x_{23} = \min\{3,7\}=3$.

	D_1	D_2	D_3	D_4	
O_1	19 5	30 2	50	10	0
O_2	70	30 6	40 3	60	0
O_3	40	8	70	20	18
	0	0	4	14	

The 2nd row has supply zero. We move to cost cell x_{33} and find min$\{18, 4\}$=4.

	D_1	D_2	D_3	D_4	
O_1	19 — 5	30 — 2	50	10	0
O_2	70	30 — 6	40 — 3	60	0
O_3	40	8	70 — 4	20	14
	0	0	0	14	

At this stage, we get equal supply and demand that is 14. Therefore, we move to cost cell x_{34} with entry 14. We now stop.

	D_1	D_2	D_3	D_4	
O_1	19 — 5	30 — 2	50	10	0
O_2	70	30 — 6	40 — 3	60	0
O_3	40	8	70 — 4	20 — 14	0
	0	0	0	0	

Finally, the initial basic feasible solution is

$$x_{11} = 5, \quad x_{12} = 2, \quad x_{22} = 6, \quad x_{23} = 3, \quad x_{33} = 4, \quad x_{34} = 14.$$

The corresponding transportation cost is given by

$$= x_{11}c_{11} + x_{12}c_{12} + x_{22}c_{22} + x_{23}c_{23} + x_{33}c_{33} + x_{34}c_{34}$$
$$= 5 \times 19 + 2 \times 30 + 6 \times 30 + 3 \times 40 + 4 \times 70 + 14 \times 20$$
$$= 1015 \text{ units.}$$

We can solve the transportation problem using MATLAB function **nwc.m**. See the following Code 8.1 for northwest corner method.

Code 8.1: nwc.m label

```
function [minTcost,b,c]=nwc(A,sup,dem)
%input : augmented matrix A, supply sup,
```

```
%demand dem
%output: cost matrix c, basis matrix B
[m,n]=size(A);
minTcost=0;
I=1;
J=1;
b=zeros(m,n);
c=zeros(m,n);
while I<m+1
        while J <n+1
                if sup(I,1)<dem(1,J)
minTcost=minTcost+(A(I,J)*sup(I,1));
c(I,J)=sup(I,1);
disp(c);
dem(1,J)=dem(1,J)-sup(I,1);
x=sprintf('x(%d,%d)=%d',I,J,sup(I,1));
disp(x);
disp('dem=');
disp(dem);
sup(I,1)=0;
disp('sup');
disp(sup);
b(I,J)=1;
I=I+1;
end
if sup(I,1)>dem(1,J)
minTcost=minTcost+(A(I,J)*dem(1,J));
c(I,J)=dem(1,J);
disp(c);
% disp(dem);
sup(I,1)=sup(I,1)-dem(1,J);
x=sprintf('x(%d,%d)=%d',I,J,dem(1,J));
disp(x);
disp('sup');
disp(sup);
dem(1,J)=0;
disp('dem');
disp(dem);
b(I,J)=1;
```

```
J=J+1;
end
if  sup(I,1)==dem(1,J)
minTcost=minTcost+(A(I,J)*dem(1,J));
c(I,J)=dem(1,J);
disp(c);
x=sprintf('x(%d,%d)=%d',I,J,dem(1,J));
disp(x);
sup(I,1)=0;
dem(1,J)=0;
disp('sup');
disp(sup);
disp('dem')
disp(dem);
b(I,J)=1;
I=I+1  ;
J=J+1  ;
end
end
end
end
```

Example 8.3. Solve the transportation problem in MATLAB.

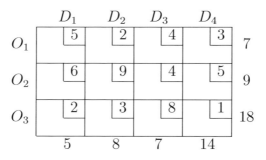

In the Command Window,

>> A= $\begin{bmatrix} 5 & 2 & 4 & 3; & 6 & 9 & 4 & 5; & 2 & 3 & 8 & 1 \end{bmatrix}$

>> sup= $\begin{bmatrix} 7; & 9; & 18 \end{bmatrix}$

>> dem= $\begin{bmatrix} 5 & 8 & 7 & 14 \end{bmatrix}$

```
>>[minTcost]=nwc(A, sup, dem)
```

Output:

```
>>
    x(1,1)=5
    x(1,2)=2
    x(2,2)=6
    x(2,3)=3
    x(3,3)=4
    x(3,4)=14

minTcost=
        141
```

Example 8.4. The cost matrix of a transportation problem is given by

1	2	3	4
4	3	2	1
0	2	2	1

The following are the values of variables in a feasible solution $x_{11} = 3, x_{12} = 6, x_{23} = 2, x_{24} = 6, x_{31} = 4, x_{33} = 6$. Then, which of the following is TRUE?

(a) The solution is degenerate and basic.

(b) The solution is nondegenerate and basic.

(c) The solution is degenerate and nonbasic.

(d) The solution is nondegenerate and nonbasic.

We have $m + n - 1 = 3 + 4 - 1 = 6$. In a balanced transportation problem, there are atmost $m + n - 1$ basic variables. None of x_{ij} is zero, therefore the solution is nondegenerate and basic. Thus, option (b) is true.

8.4   Least Cost Method

This method usually provides a better starting basic feasible solution than the northwest corner method since it takes into account the cost variables in the problem.

Algorithm

1. Allocate $\min\{a_i, b_j\}$ to the cell having lowest cost in the transportation matrix. If there is a tie, then choose arbitrarily.

2. Ignore the row or column which is satisfied. If a row and column are both satisfied, then ignore only one of them.

3. Adjust a_i and b_j for those rows and columns which are not ignored.

4. Repeat steps 1–3 until all units have been allocated.

Example 8.5. Determine an initial basic feasible solution of the following balanced transportation problem by the least cost method.

	D_1	D_2	D_3	D_4	
O_1	6	4	1	5	14
O_2	8	9	2	7	16
O_3	4	3	6	2	5
	6	10	15	4	

Supply   : $a_1 = 14, a_2 = 16, a_3 = 5$.
Demand : $b_1 = 6, b_2 = 10, b_3 = 15, b_4 = 4$.

Note that the least cost is 1 at $(1, 3)$ cell in above transportation matrix. Therefore, we allocate $\min\{14, 15\} = 14$ at (1,3) cell.

	D_1	D_2	D_3	D_4	
O_1	6	4	1 14	5	0
O_2	8	9	2	7	16
O_3	4	3	6	2	5
	6	10	1	4	

Ignore the 1st row. The least costs are at $(2,3)$ and $(3,4)$ cell. We choose any one. Allocate min $\{16,1\} = 1$ at $(2,3)$ cell in the above transportation matrix.

	D_1	D_2	D_3	D_4	
O_1	6	4	1 14	5	0
O_2	8	9	2 1	7	15
O_3	4	3	6	2	5
	6	10	0	4	

Ignore the 3rd column. The least cost is 2 at $(3,4)$ cell in transportation matrix. Therefore, we allocate min $\{5,4\} = 4$ at $(3,4)$ cell.

	D_1	D_2	D_3	D_4	
O_1	6	4	1 14	5	0
O_2	8	9	2 1	7	15
O_3	4	3	6	2 4	1
	6	10	0	0	

The least cost is 3 at $(3,2)$ cell. Allocate min $\{1,10\} = 1$ at $(3,2)$ cell. We ignore the 3rd row.

	D_1	D_2	D_3	D_4	
O_1	6	4	1 / 14	5	0
O_2	8	9	2 / 1	7	15
O_3	4 / 1	3	6	2 / 4	0
	6	9	0	0	

The least cost is 4 at $(3,1)$ cell, but the 3rd row is satisfied and ignored. Therefore, we look for the next least cost which is 6 at cell $(3,3)$, but its row and column are satisfied and ignored. Next, the least cost is 7 at $(2,4)$ cell which is also ignored. Finally, the least cost is 8 at $(2,1)$ cell. We allocate min $\{15,6\} = 6$ at $(2,1)$ cell and the 1st column is ignored.

	D_1	D_2	D_3	D_4	
O_1	6	4	1 / 14	5	0
O_2	8 / 6	9	2 / 1	7	9
O_3	4 / 1	3	6	2 / 4	0
	0	9	0	0	

This time, the least cost is 9 at $(2,2)$ cell. Therefore, we allocate min $\{9,9\} = 9$ at $(2,2)$ cell.

	D_1	D_2	D_3	D_4	
O_1	6	4	1 / 14	5	0
O_2	8 / 6	9 / 9	2 / 1	7	0
O_3	4 / 1	3	6	2 / 4	0
	0	0	0	0	

All supplies and demands are satisfied. Therefore, the least cost method is over. The initial basic feasible solution is

$$x_{13} = 14, x_{21} = 6, x_{22} = 9, x_{23} = 1, x_{32} = 1, x_{34} = 4.$$

The transportation cost is given by

$$= 1 \times 14 + 8 \times 6 + 9 \times 9 + 2 \times 1 + 3 \times 1 + 2 \times 4 = 156 \text{ units.}$$

Example 8.6. Solve the following transportation problem by the least cost method.

	D_1	D_2	D_3	D_4	
O_1	5	4	3	2	5
O_2	10	8	4	7	5
O_3	9	9	8	4	5
	1	6	2	6	

Supply : $a_1 = 5, a_2 = 5, a_3 = 5$.
Demand : $b_1 = 1, b_2 = 6, b_3 = 2, b_4 = 6$.

We identify the least cost cell as $(1,4)$ cell with cost 2. We have $a_1 = 5$ and $b_4 = 6$. We allocate $\min\{5, 6\} = 5$ and new $a_1 = 0$ and $b_4 = 6 - 5 = 1$.

	D_1	D_2	D_3	D_4	
O_1	5	4	3	2 5	0
O_2	10	8	4	7	5
O_3	9	9	8	4	5
	1	6	2	1	

Ignore the 1st row. The least cost is 4 in the two cell: $(2,3)$ and $(3,4)$. We choose $(3,4)$ cell and allocate $\min\{5,1\} = 1$. We have $a_3 = 5 - 1 = 4$ and $b_4 = 0$.

	D_1	D_2	D_3	D_4	
O_1	5	4	3	2 5	0
O_2	10	8	4	7	5
O_3	9	9	8	4 1	4
	1	6	2	0	

Ignore the 4th column. The least cost cell in the new submatrix is (2,3) with cost 4. Thus, we allocate $\min\{5,2\} = 2$ and new $a_2 = 5 - 2 = 3$ and $b_3 = 0$.

	D_1	D_2	D_3	D_4	
O_1	5	4	3	2 / 5	0
O_2	10	8	4 / 2	7	3
O_3	9	9	8	4 / 1	4
	1	6	0	0	

Ignore the 3rd column. The least cost cell is (2,2) with cost 8 in the new submatrix. We allocate $\min\{3,6\} = 3$ and new $a_2 = 0$ and $b_2 = 6 - 3 = 3$.

	D_1	D_2	D_3	D_4	
O_1	5	4	3	2 / 5	0
O_2	10	8 / 3	4 / 2	7	0
O_3	9	9	8	4 / 1	4
	1	3	0	0	

Ignore the 2nd row. The least cost cells are (3,1) and (3,2). We choose (3,2) cell arbitrarily and allocate $\min\{4,3\} = 3$. We have new $a_3 = 4 - 3 = 1$ and $b_2 = 0$.

	D_1	D_2	D_3	D_4	
O_1	5	4	3	2 / 5	0
O_2	10	8 / 3	4 / 2	7	0
O_3	9	9 / 3	8	4 / 1	1
	1	0	0	0	

Ignore the 2nd column. We have only one cell left, that is (3,1) cell. Therefore we allocate $\min\{1,1\} = 1$. We have new $a_3 = 0$ and $b_1 = 0$.

	D_1	D_2	D_3	D_4	0
O_1	5	4	3	2 (5)	0
O_2	10 (3)	8 (2)	4	7	0
O_3	9 (1)	9 (3)	8	4 (1)	0
	0	0	0	0	0

All supplies and demands are satisfied. Thus, least cost method is over and the initial basic feasible solution is

$$x_{14} = 5, x_{22} = 3, x_{23} = 2, x_{31} = 1, x_{32} = 3, x_{34} = 1.$$

The minimum transportation cost is

$$= 5 \times 2 + 3 \times 8 + 2 \times 4 + 1 \times 9 + 3 \times 9 + 1 \times 4$$
$$= 82 \text{ units.}$$

We have written MATLAB function `lcm.m` for the least cost method as given in the following Code 8.2.

Code 8.2: lcm.m

```
function  [minTcost,b,c]=leastcost(A,sup,dem)
%input:   Transportation cost A, supply sup,
%         demand dem
 %output: minimum transportation cost
%minTcost, basic matrix b, cost matrix c
   [m,n]=size(A);
   sum=0;
   rf=zeros;
   cf=zeros;
   b=zeros(m,n);
   c=zeros(m,n);
   for  I=1:m
       rf(I,1)=0;
       for  J=1:n
           cf(1,J)=0;
       end
   end
   f=m;
```

```
d=n;
while  f>0 && d>0
      min=Inf;
      for  I=1:m
            if  rf(I,1)~=1
                  for  J=1:n
                        if  cf(1,J)~=1
                              if  min>A(I,J)
                                    min=A(I,J);
                                    p=I;
                                    q=J;
                              end
                        end
                  end
            end
      end
      if  sup(p,1)<dem(1,q)
            b(p,q)=1;
            c(p,q)=sup(p,1);
            disp(c);
            sum=sum+A(p,q)*sup(p,1);
      x=sprintf('x(%d,%d)=%d',p,q,sup(p,1));
            disp(x);
            dem(1,q)=dem(1,q)-sup(p,1);
            sup(p,1)=0;
            disp('sup');
            disp(sup);
            disp('dem');
            disp(dem);
            rf(p,1)=1;
            f=f-1;
        else
            if  sup(p,1)>dem(1,q)
                  b(p,q)=1;
                  c(p,q)=dem(1,q);
                  disp(c);
                  sum=sum+A(p,q)*dem(1,q);
            x=sprintf('x(%d,%d)=%d',p,q,dem(1,q));
                  disp(x);
```

```
            sup(p,1)=sup(p,1)−dem(1,q);
            dem(1,q)=0;
            disp('sup');
            disp(sup);
            disp('dem');
            disp(dem);
            cf(1,q)=1;
            d=d−1;
        else
            if sup(p,1)==dem(1,q)
                b(p,q)=1;
                c(p,q)=sup(p,1);
                disp(c);
                sum=sum+A(p,q)*sup(p,1);
    x=sprintf('x(%d,%d)=%d',p,q,sup(p,1));
                sup(p,1)=0;
                dem(1,q)=0;
                disp('sup');
                disp(sup);
                disp('dem');
                disp(dem);
                disp(x);
                rf(p,1)=1;
                cf(1,q)=1;
                f=f−1;
                d=d−1;
            end
        end
    end
end
minTcost=sum;
return
```

Example 8.7. Solve the following transportation problem by the least cost method in MATLAB.

	D_1	D_2	D_3	D_4	
O_1	3	1	5	10	2
O_2	7	4	5	3	6
O_3	8	8	2	2	7
	3	3	4	5	

In the Command Window,

```
>> A=[3  1  5  10;  7  4  5  3;  8  8  2  2]
```

```
>> sup=[2;  6;  7]
```

```
>> dem=[3  3  4  5]
```

```
>> [minTcost]=leastcostnew(A,sup,dem)
```

Output:

```
x(1,2)=2
x(3,3)=4
x(3,4)=3
x(2,4)=2
x(2,2)=1
x(2,1)=3
```

minTcost=

47

Example 8.8. Solve by the least cost method in MATLAB.

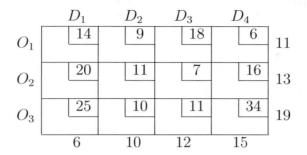

In the Command Window,

```
>> A=[14  9  18  6;  20  11  7  16;  25  10  11  34]

>> sup=[11;  13;  19]

>> dem=[6  10  12  15]

>> [minTcost]=leastcost(A,sup,dem)
```

Output:

```
x(1,4)=11
x(2,3)=12
x(3,2)=10
x(2,4)=1
x(3,1)=6
x(3,4)=3

minTcost =
          518
```

8.5   Vogel's Approximation Method

This method is preferred over the other two methods discussed above because it produces best possible minimize cost and close to an optimal solution. Therefore, if we use the starting basic feasible solution obtained by Vogel's approximation method and proceed to solve for the optimum solution, then the time required to arrive at the optimum solution is greatly reduced. W. R. Vogel developed this method.

Algorithm

1. Take the 1st row and choose its smallest cost and subtract this from the cost which is the next highest cost, and write the result in front of the row on the right. This is the penalty for the first row. In this way, compute the penalty of each row. Similarly, calculate column penalties and write those in the bottom of the cost matrix below corresponding columns.

2. Select the highest penalty and observe the row or column for which this corresponds. Then, make allocation min $\{a_i, b_j\}$ to the cell having the lowest cost in the selected row or column.

3. Ignore the row or cell which is satisfied. Calculate fresh penalties for the remaining sub-matrix as in step 1 and for allocation, follow the procedure of step 2. Continue the process untill all rows and columns are satisfied.

Rules for tie:

In case of a tie for the largest penalty, choose the lowest cost cell in all tied rows and columns for allocation. Again, if there is a tie for the lowest cost cell; select one for allocation which gives minimum $c_{ij}x_{ij}$.

Example 8.9. Consider the transportation problem:

	D_1	D_2	D_3	D_4	
O_1	19	30	50	10	7
O_2	70	30	40	60	9
O_3	40	8	70	20	18
	5	8	7	14	

Obtain a starting basic feasible solution for transportation problem using Vogel's approximation method.

As the problem is balanced, we find a starting basic feasible solution.

	D_1	D_2	D_3	D_4		row penalty
O_1	19	30	50	10	7	9
O_2	70	30	40	60	9	10
O_3	40	8	70	20	18	12
	5	8	7	14		
column penalty	21	22	10	10		

The highest penalty is 22 which is the 2nd column. Observe that the lowest cost cell in the 2nd column is 8 at (3,2) cell. We allocate min $\{18, 8\} = 8$ at (3,2) cell and ignore the 2nd column.

	D_1	D_2	D_3	D_4		row penalty
O_1	19	30	50	10	7	9
O_2	70	30	40	60	9	20
O_3	40	8 8	70	20	10	20
	5	0	7	14		
column penalty	21	×	10	10		

The highest penalty is 21 which is the 1st column. Observe that the lowest cost is 19. We allocate min $\{7, 5\} = 5$ in $(1, 1)$ cell.

	D_1	D_2	D_3	D_4		row penalty
O_1	19 5	30	50	10	2	40
O_2	70	30	40	60	9	20
O_3	40	8 8	70	20	10	50
	0	0	7	14		
column penalty	×	×	10	10		

We now compute penalties of each row and column. Note that the highest penalty is 50 in the 3rd row and the least cost is 20 in $(3, 4)$ cell. Therefore, we allocate min$\{10, 14\}$=10 at $(3, 4)$ cell. In this way, the third row is satisfied.

	D_1	D_2	D_3	D_4		row penalty
O_1	19 5	30	50	10	2	40
O_2	70	30	40	60	9	20
O_3	40	8 8	70	20 10	0	×
	0	0	7	4		
column penalty	×	×	10	50		

We again calculate penalties of remaining rows and columns. This time, the highest penalty is 50 in the 4th column and the least cost is 10 in that column. Therefore, we find $\min\{2,4\}=2$ at $(1,4)$ cell. Thus, the 1st row is satisfied.

	D_1	D_2	D_3	D_4		row penalty
O_1	19 5	30	50	10 2	0	×
O_2	70	30	40	60	9	20
O_3	40	8 8	70	20 10	0	×
	0	0	7	2		
column penalty	×	×	40	60		

The highest penalty is 60 in the 4th column and the least cost is 60. Thus, we find $\min\{9,2\}=2$ at $(2,4)$ cell and the 4th column is satisfied.

	D_1	D_2	D_3	D_4		row penalty
O_1	19 5	30	50	10 2	0	×
O_2	70	30	40	60 2	7	40
O_3	40	8 8	70	20 10	0	×
	0	0	7	0		
column penalty	×	×	40	×		

The remaining highest penalty is 40 and the least cost cell is $(2,3)$. Thus, we allocate $\min\{7,7\}=7$ at $(2,3)$ cell. The 3rd column and the 2nd row are satisfied.

	D_1	D_2	D_3	D_4		row penalty
O_1	19 / 5	30	50	10 / 2	0	×
O_2	70	30	40 / 7	60 / 2	0	×
O_3	40	8 / 8	70	20 / 10	0	×
	0	0	0	0		
column penalty	×	×	×	×		

The initial basic feasible solution is:

$$x_{11} = 5, x_{14} = 2, x_{23} = 7, x_{24} = 2, x_{32} = 8, x_{34} = 10.$$

It is nondegenerate and the total transportation cost is computed as follows:

$$= 5 \times 19 + 2 \times 10 + 7 \times 40 + 2 \times 60 + 8 \times 8 + 10 \times 20$$
$$= 779 \text{ units.}$$

Example 8.10. Calculate the initial basic feasible solution for this transportation problem by Vogel's Approximation Method.

	D_1	D_2	D_3	D_4	
O_1	2	2	1	5	300
O_2	8	2	6	5	300
O_3	6	1	4	2	200
	200	200	300	100	

We firstly calculate row penalty and column penalty.

	D_1	D_2	D_3	D_4		row penalty
O_1	2	2	1	5	300	1
O_2	8	2	6	5	300	3
O_3	6	1	4	2	200	1
	200	200	300	100		
column penalty	4	1	3	3		

The highest penalty is 4. We allocate min$\{300, 200\} = 200$ at $(1,1)$ cell because cell $(1,1)$ contains the least cost for the 1st column.

	D_1	D_2	D_3	D_4		row penalty
O_1	2 200	2	1	5	100	1
O_2	8	2	6	5	300	3
O_3	6	1	4	2	200	1
	0	200	300	100		
column penalty	×	1	3	3		

The 2nd row, and 3rd column have equal and highest penalties. Since the least cost is 1 at (1,3) cell in the 3rd column, we allocate $\min\{100, 300\} = 100$ at (1,3) cell.

	D_1	D_2	D_3	D_4		row penalty
O_1	2 200	2	1 100	5	0	×
O_2	8	2	6	5	300	3
O_3	6	1	4	2	200	1
	0	200	200	100		
column penalty	×	1	3	3		

The highest penalty is 3 and the least cost is 2. Thus, we allocate $\min\{200, 100\} = 100$ at (3,4) cell.

	D_1	D_2	D_3	D_4		row penalty
O_1	2 200	2	1 100	5	0	×
O_2	8	2	6	5	300	4
O_3	6	1	4	2 100	100	3
	0	200	200	0		
column penalty	×	1	2	×		

The highest penalty is 4 which is in the 2nd row. Therefore, we allocate $\min\{300, 200\} = 200$ at (2,2) cell. This cell contains the least cost.

	D_1	D_2	D_3	D_4		row penalty
O_1	2 200	2	1 100	5	0	×
O_2	8	2 200	6	5	100	6
O_3	6	1	4	2 100	100	4
	0	0	200	0		
column penalty	×	×	2	×		

The highest penalty is 6 in the 2nd row and least cost cell is (2,3). Therefore, we allocate $\min\{100, 200\} = 100$.

	D_1	D_2	D_3	D_4		row penalty
O_1	2 / 200	2	1 / 100	5	0	×
O_2	8	2 / 200	6 / 100	5	0	×
O_3	6	1	4	2 / 100	100	4
column penalty	0	0	100	0		
	×	×	4	×		

The highest penalty is 4 and the least cost cell is 4. Thus, we allocate $\min\{100, 100\} = 100$ at $(3, 3)$ cell.

	D_1	D_2	D_3	D_4		row penalty
O_1	2 / 200	2	1 / 100	5	0	×
O_2	8	2 / 200	6 / 100	5	0	×
O_3	6	1	4 / 100	2 / 100	0	×
column penalty	0	0	0	0		
	×	×	×	×		

Thus, the initial basic feasible solution is

$$x_{11} = 200, x_{13} = 100, x_{22} = 200, x_{23} = 100, x_{33} = 100, x_{34} = 100.$$

The total transportation cost is computed as follows:

$$= 200 \times 2 + 100 \times 1 + 200 \times 2 + 100 \times 6 + 100 \times 4 + 100 \times 2$$
$$= 2100 \text{ units.}$$

We have written MATLAB function `vogel.m` for Vogel's Approximation Method in the following Code 8.3.

Code 8.3: vogel.m

```
function [minTcost,b,c] = vogel(A,sup,dem)
%input: transportation cost A, vector supply
%        sup, vector demand dem
%output:minimum transportation cost minTcost,
%        basic matrix b, cost matrix c,
b= zeros(size(A));
ctemp = A;
[m,n]=size(A);
c=zeros(m,n);
```

```matlab
while  length(find(dem==0)) < length(dem)  ||
          length(find(sup==0)) < length(sup)
prow = sort(ctemp,1);
prow = prow(2,:) - prow(1,:); % row penalty
pcol = sort(ctemp,2);
pcol = pcol(:,2) - pcol(:,1); %column penalty
[rmax,rind] = max(prow);
[cmax,cind] = max(pcol);
disp('column penalty')
disp(prow);
disp('row penalty');
disp(pcol);
%value for allocated cell
 if rmax>cmax
    [~,mind] = min(ctemp(:,rind));
[amt,dem,sup,ctemp] =
   chkdemandsupply(dem,sup,rind,mind,ctemp);
   x=sprintf('x(%d,%d)=%d',mind,rind,amt);
   disp(x);
   b(mind,rind)=1;
   c(mind,rind)=amt;
   disp(c);
end
    if cmax>= rmax
       [~,mind] = min(ctemp(cind,:));
[amt,dem,sup,ctemp] = chkdemandsupply(dem,sup,
                         mind,cind,ctemp);
       x=sprintf('x(%d,%d)=%d',cind,mind,amt);
       disp(x);
       b(cind,mind) =1;
       c(cind,mind)=amt;
       disp(c);
       minTcost = sum(sum(c.*A));
    end
end
function [y,dem,sup,ctemp] =
       chkdemandsupply(dem,sup,ded,sud,ctem)
tempd = dem;
temps = sup;
```

```
if  tempd(ded) > temps(sud)
    temps(sud) = 0;
    tempd(ded) = dem(ded) - sup(sud);
    disp('sup');
    disp(temps);
    disp('dem');
    disp(tempd);
    y = sup(sud);
  ctem(sud,:) = inf;
end
if  tempd(ded) < temps(sud)
    tempd(ded) = 0;
    temps(sud) = sup(sud) - dem(ded);
    disp('sup');
    disp(temps);
    disp('dem');
    disp(tempd);
    y = dem(ded);
    ctem(:,ded) = inf;
end
  if  tempd(ded) == temps(sud)
      tempd(ded) = 0;
      temps(sud) = 0;
      disp('sup');
      disp(temps);
      disp('dem');
      disp(tempd);
      y = dem(ded);
    ctem(:,ded) = inf;
    ctem(sud,:) = inf;
  end
dem = tempd;
sup = temps;
ctemp = ctem;
```

Example 8.11. Solve the following transportation problem by Vogel's Approximation Method in MATLAB.

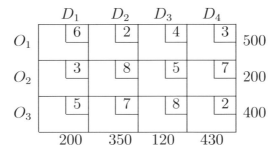

In the Command Window,

$\gg$ A=[6 2 4 3; 3 8 5 7; 5 7 8 2]

$\gg$ sup=[500; 200; 400]

$\gg$ dem=[200 350 120 430]

$\gg$ [minTcost] = vogel(A, sup, dem)

0	350	120	30
200	0	0	0
0	0	0	400

minTcost =

    2670

Example 8.12. Solve the transportation problem by Vogel's Approximation Method in MATLAB.

	D_1	D_2	D_3	D_4	
O_1	30	20	50	20	75
O_2	20	10	30	40	120
O_3	40	20	40	30	105
	65	60	80	95	

In the Command Window,

```
>> A=[30  20  50  20;  20  10  30  40;  40  20  40  30]

>> sup=[75;  120;  105]

>> dem=[65  60  80  95]

>> [minTcost,b,c] = vogel(A,sup,dem)
```

Output:

x(2,2)=60
x(1,4)=75
x(2,1)=60
x(3,1)=5
x(3,3)=80
x(3,4)=20

$\min \text{Tcost} =$

7300

Note: All the output of MATLAB functions nwc.m, lcm.m and vogel.m have not been shown in the solution of the above problems. It is left for the reader to use MATLAB to see all the output step-by-step.

8.6   Optimal Solution from BFS

We study a computational procedure to find an optimal solution from the starting basic feasible solution (BFS).

Computational Procedure

1. Introduce the variables u_i and v_j corresponding to i^{th} row and j^{th} column, respectively. Write u_i in front of each ith row and v_j at the bottom of each jth column. Take any u_i or v_j to be zero for maximum number of allocations.

2. For basic cells (which contain allocations), calculate $u_i + v_j = c_{ij}$. This relation assigns value to all u_i and v_j. Note that this relation is equivalent to $z_j - c_j = 0$ for basic variables in the simplex algorithm.

3. For nonbasic cells (which have no allocations), calculate $u_i + v_j - c_{ij}$ and write them in the southwest corner of the concerned cells of the tableau.

4. If all the southwest entries are less than or equal to zero, then the basic feasible solution is optimal. If at least one of the southwest entries is positive, then this basic feasible solution is not optimal. In this situation, look for the most positive southwest entry in the cost matrix. This decides the entering variable.

5. Assign θ (quantity in the cell having most positive southwest entry) and make a loop.

6. Start from θ cell and move horizontally and vertically to the nearest basic cell with restriction that the turn (corner) of the loop must not lie in any nonbasic cell (except θ cell). In this way, return to θ cell to complete the loop.

7. Add or subtract θ in concerned entries of the loop maintaining feasibility and value of θ is fixed as the minimum of the entries from which θ has been subtracted.

8. Inserting the fixed value of θ, we get the next basic feasible solution which improves the initial transportation cost.

9. While inserting the value θ if a cell assumes '0' value, we will not mention '0' value as this is the leaving variable, i.e., this cell has become nonbasic. Again, use the latest basic feasible solution and repeat steps 1–8 until every southwest entry turns out to be less than or equal to zero. This is an optimal solution.

Example 8.13. Find an optimal solution to the transportation problem.

	D_1	D_2	D_3	D_4	
O_1	1	2	3	4	30
O_2	7	6	2	5	50
O_3	4	3	2	7	35
	15	30	25	45	

We have applied the least cost method to find a starting basic feasible solution. It is left for the reader to solve this transportation problem step-by-step.

	D_1	D_2	D_3	D_4	
O_1	1 / 15	2 / 15	3	4	30
O_2	7	6	2 / 25	5 / 25	50
O_3	4	3 / 15	2	7 / 20	35
	15	30	25	45	

Thus, the starting basic feasible solution is $x_{11} = 15, x_{12} = 15, x_{23} = 25, x_{24} = 25, x_{32} = 15, x_{34} = 20$. It is a nondegenerate basic feasible solution. Note that we are concerned only to find the optimal solution of the transportation problem.

1. Since all the rows have the same number of allocations, therefore any of the u_1, u_2, u_3 may be assigned zero. Suppose that $u_1 = 0$

2. We allocate the values of the other dual variables using the relation: $u_i + v_j = c_{ij}$ for basic cells.
   The possible equations for the basic cells are:

$$u_1 + v_1 = 1,$$
$$u_1 + v_2 = 2,$$
$$u_2 + v_3 = 2,$$
$$u_2 + v_4 = 5,$$
$$u_3 + v_2 = 3,$$
$$u_3 + v_4 = 7.$$

Solving the above equations, we get

$$u_1 = 0, \quad u_2 = -1, \quad u_3 = 1,$$
$$v_1 = 1, \quad v_2 = 2, \quad v_3 = 3, \quad v_4 = 6.$$

Thus, $u_1, u_2, u_3, v_1, v_2, v_3, v_4$ are calculated.

3. We calculate $u_i + v_j - c_{ij}$ for each nonbasic cell and write those in the southwest corner of the cell. Note that $x_{13}, x_{14}, x_{21}, x_{22}, x_{31}, x_{33}$ are nonbasic variables. Therefore, we calculate all these values for nonbasic cells and write those in the southwest corner:

$$x_{13} = u_1 + v_3 - c_{13} = \quad 0 + 3 - 3 = \quad 0,$$
$$x_{14} = u_1 + v_4 - c_{14} = \quad 0 + 6 - 4 = \quad 2,$$
$$x_{21} = u_2 + v_1 - c_{21} = -1 + 1 - 7 = -7$$
$$x_{22} = u_2 + v_2 - c_{22} = \quad 1 + 2 - 6 = -5,$$
$$x_{31} = u_3 + v_1 - c_{31} = \quad 1 + 1 - 4 = -2,$$
$$x_{33} = u_3 + v_3 - c_{33} = \quad 1 + 3 - 2 = \quad 2.$$

	1	2	3	4
	15	15		
			0	2
	7	6	2	5
			25	25
-7		-5		
	4	3	2	7
				20
-2		15	2	

Note that all elements in the southwest corner are not less than or equal to zero. Therefore, this basic feasible solution is not optimal.

4. We look at the most positive southwest entry. This is available at $(1, 4)$ and $(3, 3)$ cells of the cost matrix. Take any one cell for the entering variable. Suppose that we take the nonbasic cell $(3, 3)$, i.e., nonbasic variable x_{33} to enter the basis.

5. We make a loop as per rule. Assign θ value to $(3,3)$ cell. Subtract and add θ at corners of the loop to maintain feasibility.

Decide the value of θ by taking the minimum of the entries from which θ have been subtracted, i.e., $\theta = \min\{25,20\} = 20$.

	1	2	3	4
15	15	0	2	
7	6	2	5	
-7	-5	25-θ ⟶ 25+θ		
4	3	2	7	
-2	15	θ ⟵ 20-θ	2	

6. The new tableau is

	1	2	3	4
	15	15	0	2
7	6	2	5	
-7	-5	5	45	
4	3	2	7	
-2		15	20	

The new basic feasible solution is $x_{11} = 15$, $x_{12} = 5$, $x_{23} = 5$, $x_{24} = 45$, $x_{32} = 15$, $x_{33} = 20$. The first iteration is over.

7. To start the second iteration, we introduce new u_1, u_2, u_3, v_1, v_2, v_3 and v_4.

The possible equations for the basic cells are

$$u_1 + v_1 = 1,$$
$$u_1 + v_2 = 2,$$
$$u_2 + v_3 = 2,$$
$$u_2 + v_4 = 5,$$
$$u_3 + v_2 = 3,$$
$$u_3 + v_3 = 2.$$

Solving the above equations, we get

$$u_1 = 0, \quad u_2 = 1, \quad u_3 = 1,$$
$$v_1 = 1, \quad v_2 = 2, \quad v_3 = 1, \quad v_4 = 4.$$

8. We calculate $u_i + v_j - c_{ij}$ for each nonbasic cell and write those in the southwest corner of the cell. Note that x_{13}, x_{14}, x_{21}, x_{22}, x_{31}, x_{34} are nonbasic variables. Therefore, we calculate all these values for nonbasic cells and write those in the southwest corner.

$$x_{13} = u_1 + v_3 - c_{13} = 0 + 1 - 3 = -2,$$
$$x_{14} = u_1 + v_4 - c_{14} = 0 + 4 - 4 = 0,$$
$$x_{21} = u_2 + v_1 - c_{21} = 1 + 1 - 7 = -5,$$
$$x_{22} = u_2 + v_2 - c_{22} = 1 + 2 - 6 = -3,$$
$$x_{31} = u_3 + v_1 - c_{31} = 1 + 1 - 4 = -2,$$
$$x_{34} = u_3 + v_4 - c_{34} = 1 + 4 - 7 = -2.$$

All elements in the southwest corner are less than or equal to zero. Therefore, the current basic feasible solution is optimal.

Thus, the optimal solution to the given transportation problem is

$$x_{11} = 15, x_{12} = 15, x_{23} = 5, x_{24} = 45, x_{32} = 15, x_{33} = 20.$$

The minimum cost of the transportation problem is

$$= 15 \times 1 + 15 \times 2 + 5 \times 2 + 45 \times 5 + 15 \times 3 + 20 \times 2$$
$$= 15 + 30 + 10 + 225 + 45 + 40$$
$$= 365 \text{ units.}$$

Example 8.14. Check the optimality of the solution obtained by Vogel's Approximation Method for the transportation problem in **Example 8.10**.

The starting basic feasible solution obtained in Example 8.10 is as follows:

	D_1	D_2	D_3	D_4
O_1	2 200	2	1 100	5
O_2	8	2 200	6 100	5
O_3	6	1	4 100	2 100

The starting transportation cost is 2100 units. The possible equations for the basic cells are

$$u_1 + v_1 = 2,$$
$$u_1 + v_3 = 1,$$
$$u_2 + v_2 = 2,$$
$$u_2 + v_3 = 6,$$
$$u_3 + v_3 = 4,$$
$$u_3 + v_4 = 2.$$

Solving the above equations, we get

$$u_1 = 0, \quad u_2 = 5, \quad u_3 = 3,$$
$$v_1 = 2, \quad v_2 = -3, \quad v_3 = 1, \quad v_4 = -1.$$

We calculate $u_i + v_j - c_{ij}$ for each nonbasic cell and write those in the southwest corner of the cell. Note that x_{12}, x_{14}, x_{21}, x_{24}, x_{31}, x_{32} are nonbasic variables. Therefore, we calculate all these values for nonbasic cells and write those in the southwest corner.

$$x_{12} = u_1 + v_2 - c_{12} = 0 - 3 - 2 = -5,$$
$$x_{14} = u_1 + v_4 - c_{14} = 0 - 1 - 5 = -6,$$
$$x_{21} = u_2 + v_1 - c_{21} = 5 + 2 - 8 = -1,$$
$$x_{24} = u_2 + v_4 - c_{24} = 5 - 1 - 5 = -1,$$
$$x_{31} = u_3 + v_1 - c_{31} = 3 + 2 - 6 = -1,$$
$$x_{32} = u_3 + v_2 - c_{32} = 3 - 3 - 1 = -1.$$

	2		2		1		5
200				100			
		-5				-6	
	8		2		6		5
		200		100			
-1						-1	
	6		1		4		2
				100		100	
-1		-1					

All elements in the southwest corner entries are less than or equal to zero. Therefore, the current basic feasible solution is optimal.

MATLAB function `multipliers2.m` is written in the following Code 8.4 to calculate the values of dual variables.

Code 8.4: multipliers2.m

```
function [u,v,b,c]=multipliers2(b,A,c,i,j)
%input : basic matrix b, transportation
%         matrix A, cost matrix c, row i,
%         column j
%output: vector u, vector v
[m,n]=size(A);
if sum(sum(b))<m+n-1
    disp('Degenerate');
else
    disp('Nondegenerate');
end
    u=Inf*ones(m,1);
    v=Inf*ones(1,n);
```

```
if ( j==0)
    u( i ,1)=0;
else
    v ( 1 , j )=0;
end
    for  row=1:m
        for  col=1:n
            if  b(row , col)>0
        if (u(row,1)~= Inf ) && ( v ( 1 , col)==Inf )
            v ( 1 , col)=A( row , col)−u( row , 1 );
            else
        if  (u(row,1)== Inf ) && ( v ( 1 , col)~= Inf )
                u( row ,1)=A( row , col)−v ( 1 , col );
                    end
                end
            end
        end
    end
for  row=1:m
    for  col=1:n
        if  b(row , col)>0
        if (u(row,1)~= Inf ) && ( v ( 1 , col)==Inf )
            v ( 1 , col)=A( row , col)−u( row , 1 );
            else
        if  (u(row,1)== Inf ) && ( v ( 1 , col)~= Inf )
                u( row ,1)=A( row , col)−v ( 1 , col );
                    end
                end
            end
        end
    end
end
for  row=1:m
    for  col=1:n
        if  b(row , col)>0
        if (u(row,1)~= Inf ) && ( v ( 1 , col)==Inf )
            v ( 1 , col)=A( row , col)−u( row , 1 );
            else
        if  (u(row,1)== Inf ) && ( v ( 1 , col)~= Inf )
                u( row ,1)=A( row , col)−v ( 1 , col );
```

```
                      end
                  end
              end
          end
      end
      for  row=1:m
          for  col=1:n
              if  b(row,col)>0
                if(u(row,1)~=Inf) && (v(1,col)==Inf)
                  v(1,col)=A(row,col)−u(row,1);
                else
              if  (u(row,1)==Inf) && (v(1,col)~=Inf)
                  u(row,1)=A(row,col)−v(1,col);
                      end
                  end
              end
          end
      end
   return
```

Note that the value of i is for the most number of allocations in a particular row, and the value of j is for the most number of allocations in a particular column in the above MATLAB function `multipliers2.m`. If i is selected, then the value of j will be zero and vice versa.

MATLAB function `uvx3.m` is written in the following Code 8.5 to calculate $u_i + v_j - c_{ij}$ for each nonbasic cell of the transportation problem.

Code 8.5: uvx3.m

```
function  [x]=uvx3(b,u,v,A)
%input  :  basic  matrix  b,  vector  u  and  v,
%          transportation  matrix  A
%output:  nonbasic  matrix  x
[m,n]=size(A);
x=zeros(m,n);
nr=1;
while  nr<m+n
    for  row=1:m
        for  col=1:n
```

```
                if  b(row,col)~=1
    x(row,col)=u(row,1)+v(1,col)-A(row,col);
                if  x(row,col)<=0
                    nr=nr+1;
                end
            end
        end
    end
end
[m,n]=size(A);
count=0;
for  I=1:m
    for  J=1:n
            if  x(I,J)<=0
                count=count+1;
            end
    end
end
if  count==(m*n)
    disp('optimality reached');
end
return
```

MATLAB function `mostpositive4.m` is written in the following Code 8.6 to choose the most positive entry of the transportation problem.

Code 8.6: mostpositive4.m

```
function  [basic,row,col]=mostpositive4(A,x,c)
%input  :  transportation  matrix  A,  nonbasic
%          matrix  x,  cost  matrix  c
%output:  element  basic,  position  row  and
%          column
    [m,n]=size(x);
    basic=0;
    count=0;
    opt=0;
    for  I=1:m
        for  J=1:n
            if  x(I,J)<0  ||  x(I,J)==0
```

```
                count=count+1;
                if  c(I,J)~=0
                    opt=opt+A(I,J)*c(I,J);
                end
            end
            if  count==(m*n)
                row=0;
                col=0;
                display('optimal cost obtained');
                x=sprintf('%d',opt);
                disp(x);
                break;
            else
                if  basic<x(I,J)
                    basic=x(I,J);
                    row=I;
                    col=J;
                end
            end
        end
    end
end
return
```

MATLAB function `cycle5.m` is written in the following Code 8.7 to make a loop of the transportation problem.

Code 8.7: cycle5.m

```
function [y,bout]=cycle5(c,row,col,b)
%input :cost matrix c, position row, col,
%        basic matrix b
%output:loop matrix y,bout
format short
bout=b;
y=c;
[m,n]=size(c);
loop=[row col];
c(row,col)=Inf;
b(row,col)=Inf;
rowsearch=1;
while (loop(1,1)~=row || loop(1,2)~=col ||
```

```
                                      length(loop)==2)
if rowsearch
    j=1;
while rowsearch
if (b(loop(1,1),j)~=0) && (j~=loop(1,2))
    loop=[loop(1,1) j;loop];
    rowsearch=0;
elseif j==n
b(loop(1,1),loop(1,2))=0;
loop=loop(2:length(loop),:);
rowsearch=0;
else
j=j+1;
end
end
else
i=1;
while ~rowsearch
if (b(i,loop(1,2))~=0)&&(i~=loop(1,1))
    loop=[i loop(1,2);loop];
    rowsearch=1;
elseif i==m
    b(loop(1,1),loop(1,2))=0;
    loop=loop(2:length(loop),:);
    rowsearch=1;
else
    i=i+1;
end
end
end
end
l=length(loop);
theta=Inf;
minindex=Inf;
for i=2:2:1
    if c(loop(i,1),loop(i,2))<theta
        theta=c(loop(i,1),loop(i,2));
        minindex=i;
    end
```

```
end
y(row,col)=theta;
for i=2:l-1
y(loop(i,1),loop(i,2))=y(loop(i,1),
                loop(i,2))+(-1)^(i-1)*theta;
end
end
```

MATLAB function `basiccell6.m` is written in the following Code 8.8 to find a basic cell from nonbasic cells.

Code 8.8: basiccell6.m

```
function [c,b,min]=basiccell6(c,y,b,row,col)
%input : cost matrix c, basic matrix b
%output: minimum value min
        [m,n]=size(c);
        min=100;
        for I=1:m
            for J=1:n
                if y(I,J)==-Inf
                    if c(I,J)<min
                        min=c(I,J);
                    end
                end
            end
        end
        for I=1:m
            for J=1:n
                if y(I,J)==-Inf
                    c(I,J)=c(I,J)-min;
                    if c(I,J)==0
                        b(I,J)=0;
                    end
                elseif y(I,J)==Inf
                    c(I,J)=c(I,J)+min;
                    b(I,J)=1;
                end
            end
```

```
    end
    c(row, col)=min;
    b(row, col)=1;
end
```
end

Example 8.15. Find the optimal transportation cost for the following transportation problem in MATLAB.

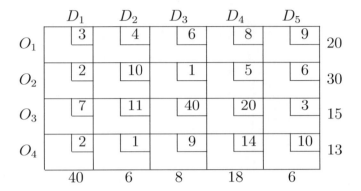

In the Command Window,

```
>> A=[3 4 6 8  9; 2 10 1  5  6;  7 11 40 20 3;
2  1 9 14 10]
```

```
>> sup=[20;  30;  15;  13]
```

```
>> dem=[40  6  8  18  6]
```

```
>> [minTcost,b,c] = vogel(A,sup,dem)
```
Output:

```
x(2,3)=8
x(3,5)=6
x(3,1)=9
x(2,1)=22
x(1,4)=18
x(4,2)=6
x(1,1)=2
x(4,1)=7
```

minTcost=
                303
b =
       1   0   0   1   0
       1   0   1   0   0
       1   0   0   0   1
       1   1   0   0   0

c =
       2   0   0   18   0
      22   0   8   0   0
       9   0   0   0   6
       7   6   0   0   0

`>> [u,v,b,c]=multipliers2(b,A,c,0,1)`

Output:

Nondegenerate

u =

       3
       2
       7
       2

v =

       0   −1   −1   5   −4

b =

       1   0   0   1   0
       1   0   1   0   0
       1   0   0   0   1
       1   1   0   0   0

c =

2	0	0	18	0
22	0	8	0	0
9	0	0	0	6
7	6	0	0	0

```
>> x=uvx3(b,u,v,A)
```

Output:

x =

0	−2	−4	0	−10
0	−9	0	2	−8
0	−5	−34	−8	0
0	0	−8	−7	−12

```
>> [basic,row,col]=mostpositive4(A,x,c)
```

basic =

2

row =

2

col =

4

```
>> [y,bout]=cycle5(c,row,col,b)
```

Output:

y =

Inf	0	0	−Inf	0
−Inf	0	8	Inf	0
9	0	0	0	6
7	6	0	0	0

bout =

1	0	0	1	0
1	0	1	0	0
1	0	0	0	1
1	1	0	0	0

```
>> [c,b,min]=basiccell6(c,y,b,row,col)
```

Output:

c =

20	0	0	0	0
4	0	8	18	0
9	0	0	0	6
7	6	0	0	0

b =

1	0	0	0	0
1	0	1	1	0
1	0	0	0	1
1	1	0	0	0

min =

    18

```
>> j=1    >> i=0

>> [u,v,b,c]=multipliers2(b,A,c,i,j)
```

Output:

Nondegenerate

u =

    3

```
        2
        7
        2

v =

        0      -1      -1       3      -4

b =

        1       0       0       0       0
        1       0       1       1       0
        1       0       0       0       1
        1       1       0       0       0

c =

       20       0       0       0       0
        4       0       8      18       0
        9       0       0       0       6
        7       6       0       0       0
>> x=uvx3(b,u,v,A)
```

Output:

optimality reached

```
x =

0      -2      -4      -2     -10
0      -9       0       0      -8
0      -5     -34     -10       0
0       0      -8      -9     -12
>> [basic,row,col]=mostpositive4(A,x,c)
```

Output:

optimal cost obtained

267

basic =

         0

row =

         0

col =

         0

8.7 Exercises

Exercise 8.1. Solve the following transportation problems using the northwest corner method.

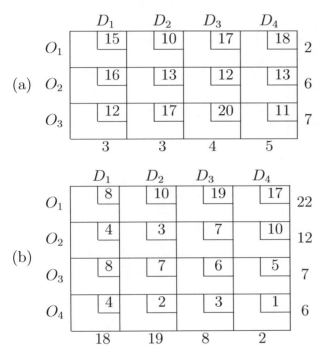

Exercise 8.2. Solve the following transportation problems using the least cost method.

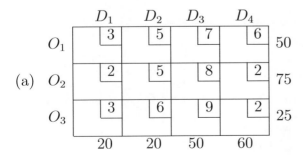

2	7	4	5
3	4	7	8
5	3	1	7
1	6	2	14
7	9	18	

(b)

Exercise 8.3. Solve the following transportation problems using Vogel's Approximation Method.

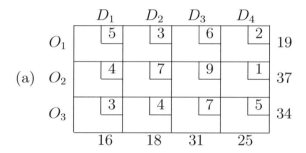

(a)

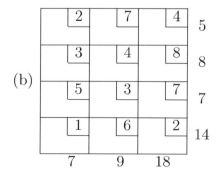

(b)

Exercise 8.4. Determine an initial basic feasible solution to the following transportation problem using the

(a) northwest corner method;

(b) least cost method;

(c) Vogel's Approximation Method.

	D_1	D_2	D_3	D_4	
O_1	10	22	10	20	8
O_2	15	20	12	8	13
O_3	20	12	10	15	11
	5	11	8	8	

Exercise 8.5. Obtain an optimal transportation cost for the following transportation problem.

	D_1	D_2	D_3	D_4	
O_1	1	2	1	4	30
O_2	3	3	2	1	50
O_3	4	2	5	9	20
	20	40	30	10	

Exercise 8.6. Determine an optimal basic feasible solution and the minimum total cost for the following transportation problem.

	D_1	D_2	D_3	D_4	D_5	
O_1	2	11	10	3	7	4
O_2	1	4	7	2	1	8
O_3	3	9	8	4	12	9
	3	3	4	5	6	

Exercise 8.7. Find an optimal solution for and the corresponding cost of the transportation problem.

	D_1	D_2	D_3	D_4	
O_1	6	1	9	3	70
O_2	11	5	2	8	55
O_3	10	12	4	7	90
	85	35	50	45	

Exercise 8.8.   Find an optimal solution for the following minimization transportation problem.

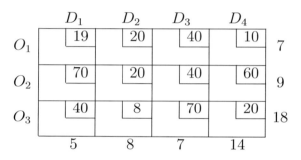

	D_1	D_2	D_3	D_4	
O_1	19	20	40	10	7
O_2	70	20	40	60	9
O_3	40	8	70	20	18
	5	8	7	14	

Chapter 9

The Assignment Problem

9.1 Introduction

An assignment problem is a particular case of the transportation problem. The goal of the assignment problem is to minimize the cost or time to finish the number of jobs assigned to the number of persons. An important characteristic of the assignment problem is that the number of jobs is equal to the number of persons.

The mathematical model of the assignment problem is given as

$$x_{ij} = \begin{cases} 1 & \text{if } i^{\text{th}} \text{ job is assigned to } j^{\text{th}} \text{ person,} \\ 0 & \text{if } i^{\text{th}} \text{ job is not assigned to } j^{\text{th}} \text{ person.} \end{cases}$$

Suppose that c_{ij} is the cost if i^{th} job is assigned to j^{th} person. Our goal is to minimize the cost c_{ij} associated with an assignment x_{ij}, then the linear programming problem can be formulated for the assignment problem as

$$
\begin{aligned}
\text{minimize} \quad & \sum_{i=1}^{m} \sum_{j=1}^{m} c_{ij} x_{ij} \\
\text{subject to} \quad & \sum_{i=1}^{m} x_{ij} = 1, \\
& \sum_{j=1}^{m} x_{ij} = 1, \\
& x_{ij} = 0 \text{ or } 1, \quad 1 \leq i, j \leq m.
\end{aligned}
\tag{9.1}
$$

The first set of constraints in (9.1) indicate that for each j, there is exactly one i for which $x_{ij}=1$, that is, for each job there is exactly one person. Similarly, the second set of constraints in (9.1) indicate that for each i, there is exactly one j for which $x_{ij} = 1$, that is, each person is assigned exactly one job.

Note that a transportation problem reduces to an assignment problem, if $m = n$ and $a_i = b_j$, where $i, j = 1, 2, \ldots, m$.

Theorem 9.1. *If a constant is added or subtracted to every element of a row and/or column of the cost matrix of an assignment problem, then the resulting assignment problem has the same optimal solution as the original problem.*

Proof. Mathematically, the theorem can be stated as

If $z = \sum\limits_{i=1}^{m} \sum\limits_{j=1}^{m} c_{ij} x_{ij}$ over all x_{ij} such that $\sum\limits_{i=1}^{m} x_{ij} = \sum\limits_{j=1}^{m} x_{ij} = 1$, $x_{ij} \geq 0$,

then $x_{ij} = x_{ij}^*$ and also minimizes $z^* = \sum\limits_{i=1}^{m} \sum\limits_{j=1}^{m} c_{ij}^* x_{ij}$ where $c_{ij}^* = c_{ij} \pm u_i \pm v_j$, for $i, j = 1, 2, \ldots, m$ and u_i, and v_j are real numbers. We now prove it. We know that

$$z^* = \sum_{i=1}^{m} \sum_{j=1}^{m} c_{ij}^* x_{ij}$$

$$= \sum_{i=1}^{m} \sum_{j=1}^{m} (c_{ij} \pm u_i \pm v_j) x_{ij}$$

$$= \sum_{i=1}^{m} \sum_{j=1}^{m} c_{ij} x_{ij} \pm \sum_{i=1}^{m} \sum_{j=1}^{m} u_i x_{ij} \pm \sum_{i=1}^{m} \sum_{j=1}^{m} v_j x_{ij}$$

$$= z \pm \sum_{i=1}^{m} u_i \sum_{j=1}^{m} x_{ij} \pm \sum_{j=1}^{m} v_j \sum_{i=1}^{m} x_{ij}.$$

Since

$$\sum_{i=1}^{m} x_{ij} = \sum_{j=1}^{m} x_{ij} = 1,$$

therefore

$$z^* = z \pm \sum_{i=1}^{m} u_i \pm \sum_{j=1}^{m} v_j.$$

Note that z^* is minimum when z is minimum. Thus, $x_{ij} = x_{ij}^*$ also minimizes z^*. □

Theorem 9.2. *If all $c_{ij} \geq 0$ and there exists a solution $x_{ij} = x_{ij}^*$ such that $z^* = \sum\limits_{i=1}^{m} \sum\limits_{j=1}^{m} c_{ij}^* x_{ij} = 0$, then the present solution is an optimal solution.*

Proof. Given that $c_{ij} \geq 0$, therefore, the value of $z = \sum\limits_{i=1}^{m} \sum\limits_{j=1}^{m} c_{ij} x_{ij}$ cannot be negative. Thus, its minimum value is zero which is attained at $x_{ij} = x_{ij}^*$. Therefore, the present solution is optimal. $\square$

Definition 9.1 (Reduced Matrix). If a matrix contains at least one zero in each row and column, then such matrix is called a "reduced matrix".

9.2 Hungarian Method

The Hungarian method was developed in 1955 by H. Kuhn. He gave it the name "Hungarian Method" because two Hungarian mathematicians, Denes Konig and Jeno Egervary, had already worked on this algorithm.

Algorithm

1. Choose the smallest cost entry in each row of the tableau, subtract this smallest cost entry from each entry in that row of tableau to get the reduced matrix.

2. If the tableau is a reduced matrix, then go to step 3, otherwise do the same procedure of step 1 for the columns which do not have at least one zero.

3. For the first assignment, choose the row having only one zero. Box this zero and cross all other zeros of the row and column in which this boxed zero lies.

4. If each zero of the reduced matrix is either boxed or crossed, and each row and column contains exactly one boxed zero, then optimality is reached so stop, otherwise go to step 5.

5. Draw the minimum number of horizontal and vertical lines so that all the zeros are covered. This can be done easily by first covering zeros in that row or column which has the maximum number of zeros. In case of a tie, we take any one, and search the rows or columns having the next lower number of zeros. Continue this until all zeros are covered.

6. Locate the smallest entry from the uncovered entries, say 'x'. Subtract 'x' from all entries not covered by these lines and add 'x' to all those entries that lie at the intersection of these lines. The entries lying on these lines but not on the intersection must be left unchanged.

7. Go to step 3.

Note: For application of the above algorithm, we assume that all c_{ij} are non-negative and the assignment problem is of the minimization case.

Example 9.1. Four persons A, B, C, D are assigned to work on four different machines I, II, III, IV. The following table shows how long it takes for a specific person to finish a job at a specific machine.

		(*Machine*)			
		I	II	III	IV
	A	8	26	17	11
(*Person*)	B	13	24	4	26
	C	38	15	18	15
	D	19	22	14	10

Find the optimal solution, i.e., how the machines should be assigned to A, B, C, D so that the job could be completed in minimum time.

We want to find $x_{ij} = 0$ or 1 so as to minimize $z = \sum_{i=1}^{4} \sum_{j=1}^{4} c_{ij} x_{ij}$.

1. We subtract the smallest entry of each row from all the entries

of the respective rows.

	I	II	III	IV
A	0	18	9	3
B	9	20	0	22
C	23	0	3	0
D	9	12	4	0

We achieved zero in each row and column. Therefore, the matrix is a reduced matrix.

2. For the first assignment, choose the row having only one zero and box this zero and cross all other zeros of the row and column in which the boxed zero lies.

	I	II	III	IV
A	[0]	18	9	3
B	9	20	[0]	22
C	23	[0]	3	⌀
D	9	12	4	[0]

3. Each zero of the reduced matrix is either boxed or crossed. Since each row and column contains exactly one boxed zero, therefore optimality is reached.

   The optimal assignment is

   $$A \to I, B \to III, C \to II, D \to IV.$$

   The optimal value is

   $$8 + 4 + 15 + 10 = 37.$$

Example 9.2. Consider the following assignment problem:

(*Job*)

		I	II	III	IV	V
	A	5	5	7	4	8
	B	6	5	8	3	7
(*Person*)	C	6	8	9	5	10
	D	7	6	6	3	6
	E	6	7	10	6	11

Find the optimal solution to minimize the total time.

1. Row reduction

1	1	3	0	4
3	2	5	0	4
1	3	4	0	5
4	3	3	0	3
0	1	4	0	5

2. Column reduction

1	0	0	0	1
3	1	2	0	1
1	2	1	0	2
4	2	0	0	0
0	0	1	0	2

3. Each row and column do not have a boxed zero. Thus, the optimal solution is not obtained.

1	[0]	0̸	0̸	1
3	1	2	[0]	1
1	2	1	0̸	2
4	2	[0]	0̸	0̸
[0]	0̸	1	0̸	2

4. We draw the minimum number of horizontal and vertical lines so that all the zeros are covered. Each horizontal line must pass through an entire row and each vertical line must pass through an entire column. This can be done easily by first covering zeros in that row or column which has a maximum number of zeros. We search the rows or columns having the next lower number of zeros and continue this till all zeros are covered. Note that only four lines are required to cover all zeros, i.e., only four assignments could be made at this stage.

1	0	0	0	1
3	1	2	0	1
1	2	1	0	2
4	2	0	0	0
0	0	1	0	2

5. The smallest uncovered entry is 1. We subtract 1 from all elements not covered by these lines and add 1 to all those elements that lie at the intersection of these lines. Note that the entries lying on these lines, but not on the intersection must be left unchanged.

1	0	0	1	1
2	0	1	0	0
0	1	0	0	1
4	2	0	1	0
0	0	1	1	2

6. No single zero is in any row or column of the tableau. Therefore, we should go for rows with two zeros. Cell $(1, 2)$ contains the first zero in the 1st row. We box it and cross all other zeros in the 1st row and 2nd column. We move row wise. There is a single 0 in the 5th row at cell $(5, 1)$. We apply the same process to box this and move column-wise until all zeros are either boxed or crossed.

1	[0]	⨉	1	1
2	⨉	1	[0]	⨉
⨉	1	[0]	⨉	1
4	2	⨉	1	[0]
[0]	⨉	1	1	2

Each row and column contains exactly one boxed zero. Therefore, optimality is reached.

The optimal assignment is

Person	Job	Time
A	II	5
B	IV	3
C	III	9
D	V	6
E	I	6
		29

Example 9.3. Solve the following assignment problem.

(*Machine*)

		I	II	III	IV	V
	A	4	6	5	1	2
	B	6	9	9	7	4
(*Man*)	C	5	8	5	5	1
	D	1	3	3	2	1
	E	6	8	7	6	2

We simply follow the following procedure:

1. We reduce the matrix row-wise.

	I	II	III	IV	V
A	3	5	4	0	1
B	2	5	5	3	0
C	4	7	4	4	0
D	0	2	2	1	0
E	4	6	5	4	0

2. Note that 2nd and 3rd columns do not contain zero. Therefore,

we reduce the matrix column-wise.

	I	II	III	IV	V
A	3	3	2	0	1
B	2	3	3	3	0
C	4	5	2	4	0
D	0	0	0	1	0
E	4	4	3	4	0

3. Each row and column do not have a boxed zero.

	I	II	III	IV	V
A	3	3	2	[0]	1
B	2	3	3	3	[0]
C	4	5	2	4	0̸
D	[0]	0̸	0̸	1	0̸
E	4	4	3	4	0̸

4. We draw the minimum number of horizontal and vertical lines to supress all zeros.

3	3	2	0	1
2	3	3	3	0
4	5	2	4	0
0	0	0	1	0
4	4	3	4	0

Number of lines=3< number of rows=5.

5. The smallest entry is 2 from the uncovered lines. We subtract it from the uncovered entries and add it to entries on the

intersection of the lines.

	I	II	III	IV	V
A	1	1	0	0	1
B	0	1	1	3	0
C	2	3	0	4	0
D	0	0	0	3	2
E	2	2	1	4	0

6. Each row and column have a boxed zero. Therefore, we reached to the optimal solution.

	I	II	III	IV	V
A	1	1	ø̶	[0]	1
B	[0]	1	1	3	ø̶
C	2	3	[0]	4	ø̶
D	ø̶	[0]	ø̶	3	2
E	2	2	1	4	[0]

The optimal assignment is

$$A \to IV, B \to I, C \to III, D \to II, E \to V$$

and the optimal value is

$$1 + 6 + 5 + 3 + 2 = 17.$$

Example 9.4. A group of five men and five women live on an island. The amount of happiness that i^{th} man and j^{th} woman derive by spending a fraction x_{ij} of their lives together is $c_{ij}x_{ij}$, where c_{ij} is given in the table below:

		(woman)			
	W_1	W_2	W_3	W_4	W_5
M_1	4	2	4	5	2
M_2	4	5	4	1	3
M_3	4	4	3	3	5
M_4	2	2	6	4	5
M_5	3	5	7	5	2

(man)

We simply follow the following procedure:

1. According to the Hungarian Method, we subtract the minimum of each row from all the entries of the respective rows.

(*woman*)

	W_1	W_2	W_3	W_4	W_5
M_1	2	0	2	3	0
M_2	3	4	3	0	2
(*man*) M_3	1	1	0	0	2
M_4	0	0	4	2	3
M_5	1	3	5	3	0

2. Note that the 2nd and 5th rows contain a single zero, box the zeros and cross all other zeros of the column in which the boxed zero lies. Apply the same procedure for the remaining rows.

(*woman*)

	W_1	W_2	W_3	W_4	W_5
M_1	2	[0]	2	3	$\cancel{0}$
M_2	3	4	3	[0]	2
(*man*) M_3	1	1	[0]	$\cancel{0}$	2
M_4	[0]	$\cancel{0}$	4	2	3
M_5	1	3	5	3	[0]

Each row and column has a boxed zero. Therefore, we reached the optimal solution.

The optimal assignment is

$$M_1 \to W_2, M_2 \to W_4, M_3 \to W_3, M_4 \to W_1, M_5 \to W_5$$

and the optimal value is

$$2 + 1 + 3 + 2 + 2 = 10.$$

Example 9.5. Consider the assignment problem shown below. In this problem, five different optimization problems are assigned to five different researchers such that the total processing time is minimized. The matrix entries represent processing time in hours.

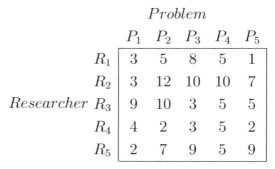

Problem

	P_1	P_2	P_3	P_4	P_5
R_1	3	5	8	5	1
R_2	3	12	10	10	7
Researcher R_3	9	10	3	5	5
R_4	4	2	3	5	2
R_5	2	7	9	5	9

1. Row reduction

2	4	7	4	0
0	9	7	7	4
6	7	0	2	2
2	0	1	3	0
0	5	7	3	7

2. Column reduction

2	4	7	2	0
0	9	7	5	4
6	7	0	0	2
2	0	1	1	0
0	5	7	1	7

3. Each row and column does not have a boxed zero.

2	4	7	2	[0]
[0]	9	7	5	4
6	7	[0]	✗	2
2	[0]	1	1	✗
✗	5	7	1	7

4. We now draw the minimum number of horizontal or vertical lines to suppress all the zeros.

2	4	7	2	0
0	9	7	5	4
6	7	0	0	2
2	0	1	1	0
0	5	7	1	7

The number of lines drawn$= 4 <$ number of rows $(= 5)$.

5. The smallest uncovered entry is 1. We subtract it from all the uncovered entries and add it to the entries on the intersection of the lines.

2	4	6	1	0
0	9	6	4	4
7	8	0	0	3
2	0	0	0	0
0	5	6	0	7

6. The number of boxed zeros is equal to the number of rows. We achieved the optimal assignment.

2	4	6	1	[0]
[0]	9	6	4	4
7	8	[0]	ø	3
2	[0]	ø	ø	ø
ø	5	6	[0]	7

The optimal assignment is

Researcher	Problem	Time
R_1	P_5	1
R_2	P_1	3
R_3	P_3	3
R_4	P_2	2
R_5	P_4	5
		14

The total processing time is 14 hours.

MATLAB function `hungarian.m` is written in the following Code 9.1 to find an optimal solution of the assignment problem.

Code 9.1: hungarian.m

```
function  [C,T]=hungarian(A)
%[C,T]=hungarian(A)
%A − a square cost matrix
%C − the optimal assignment
%T − the cost of the optimal assignment

[m,n]=size(A);
if (m~=n)
error('HUNGARIAN: Cost matrix must be
                            square!');
end
% Save original cost matrix.
orig=A;

% Reduce matrix.
A=hminired(A);

% Do an initial assignment.
[A,C,U]=hminiass(A);

% Repeat while we have unassigned rows
while (U(n+1))
% Start with no path, no unchecked zeros,
% and no unexplored rows.
LR=zeros(1,n);
LC=zeros(1,n);
CH=zeros(1,n);
RH=[zeros(1,n)  −1];

% No labelled columns.
SLC=[];

% Start path in first unassigned row
```

```
r=U(n+1);
% Mark row with end−of−path label
LR(r)=−1;
% Insert row first in labelled row set
SLR=r ;

% Repeat until we manage to find an
% assignable zero.
while (1)
% If there are free zeros in row r
if (A(r,n+1)~=0)
% ... get column of first free zero.
l=−A(r,n+1);

% If there are more free zeros in row r and
%row r is not
% yet marked as unexplored..
if (A(r,l)~=0 & RH(r)==0)
% Insert row r first in unexplored list
RH(r)=RH(n+1);
RH(n+1)=r ;

% Mark in which column the next unexplored
%zero in this row is
CH(r)=−A(r,l);
end
else
% If all rows are explored.
if (RH(n+1)<=0)
% Reduce matrix.
[A,CH,RH]=hmreduce(A,CH,RH,LC,LR,SLC,SLR);
end

% Re−start with first unexplored row
r=RH(n+1);
% Get column of next free zero in row r
l=CH(r);
% Advance 'column of next free zero '
CH(r)=−A(r,l);
```

```
% If this zero is last in the list
if (A(r,1)==0)
% ...remove row r from unexplored list
RH(n+1)=RH(r);
RH(r)=0;
end
end

% While the column l is labelled, i.e.
%in path
while (LC(l)~=0)
% If row r is explored
if (RH(r)==0)
% If all rows are explored
if (RH(n+1)<=0)
% Reduce cost matrix
[A,CH,RH]=hmreduce(A,CH,RH,LC,LR,SLC,SLR);
end

% Re-start with first unexplored row
r=RH(n+1);
end

% Get column of next free zero in row r
l=CH(r);

% Advance "column of next free zero"
CH(r)=-A(r,l);

% If this zero is last in list
if (A(r,l)==0)
% remove row r from unexplored list
RH(n+1)=RH(r);
RH(r)=0;
end
end

% If the column found is unassigned
if (C(l)==0)
```

```
% Flip all zeros along the path in LR,LC
[A,C,U]=hmflip(A,C,LC,LR,U,l,r);
%and exit to continue with next unassigned
%row.
break;
else
% ... else add zero to path

% Label column l with row r
LC(l)=r;

% Add l to the set of labelled columns
SLC=[SLC l];

% Continue with the row assigned to column l
r=C(l);

% Label row r with column l.
LR(r)=l;

% Add r to the set of labelled rows.
SLR=[SLR r];
end
end
end

% Calculate the total cost.
T=sum(orig(logical(sparse(C,1:size(orig,2),1
                                  )))));


function A=hminired(A)
%HMINIRED Initial reduction of cost
%matrix for the Hungarian method.
%B=assredin(A)
%A-the unreduced cost matrix.
%B-the reduced cost matrix with linked
%zeros in each row.
```

```
[m,n]=size(A);

%Subtract column-minimum values from each
%column
colMin=min(A);
A=A-colMin(ones(n,1),:);

%Subtract row-minimum values from each row
rowMin=min(A')';
A=A-rowMin(:,ones(1,n));

% Get positions of all zeros.
[i,j]=find(A==0);

% Extend A to give room for row zero list
% header column.
A(1,n+1)=0;
for k=1:n
% Get all column in this row.
cols=j(k==i)';
% Insert pointers in matrix.
A(k,[n+1 cols])=[-cols 0];
end


function [A,C,U]=hminiass(A)
%HMINIASS Initial assignment of the
%Hungarian method.
%[B,C,U]=hminiass(A)
%A-the reduced cost matrix.
%B-the reduced cost matrix, with assigned
% zeros removed from lists.
%C-a vector. C(J)=I means row I is
%assigned to column J,
%i.e.there is an assigned zero in
% position I,J
%U-vector with a linked list of
%unassigned rows
```

```
[n,np1]=size(A);

% Initialize  return  vectors.
C=zeros(1,n);
U=zeros(1,n+1);

% Initialize  last/next  zero  "pointers"
LZ=zeros(1,n);
NZ=zeros(1,n);

 for  i=1:n
% Set  j  to  first  unassigned  zero  in  row  i
lj=n+1;
j=A(i,lj);

% Repeat  until  we  have  no  more  zeros  (j==0)
% or  we    find  a  zero
% in  an  unassigned  column  (c(j)==0)

 while  (C(j)~=0)
% Advance  lj  and  j  in  zero  list.
lj=j;
j=A(i,lj);

% Stop  if  we  hit  end  of  list
 if  (j==0)
break;
end
end

 if  (j~=0)
% We found  a  zero  in  an  unassigned  column.

% Assign  row  i  to  column  j.
C(j)=i;

% Remove  A(i,j)  from  unassigned  zero  list
A(i,lj)=A(i,j);
```

```
% Update next/last unassigned zero pointers
NZ(i)=-A(i,j);
LZ(i)=lj;

% Indicate A(i,j) is an assigned zero
A(i,j)=0;
else
%We found no zero in an unassigned column

%Check all zeros in this row.

lj=n+1;
j=-A(i,lj);

% Check all zeros in this row for a suitable
% zero in another row.
while (j~=0)
%Check the zero in the row assigned to this
%column
r=C(j);

% Pick up last/next pointers.
lm=LZ(r);
m=NZ(r);

%Check all unchecked zeros in free list of
% this row.
while (m~=0)
% Stop if we find an unassigned column.
if (C(m)==0)
break;
end

% Advance one step in list.
lm=m;
m=-A(r,lm);
end

if (m==0)
```

```
% We failed on row r. Continue with next
% zero on row i
 lj=j ;
 j=-A( i , lj );
 else
% We found a zero in an unassigned column

%Replace zero at (r,m) in unassigned list
%  with zero at (r,j)
 A( r , lm)=-j ;
 A( r , j)=A( r ,m);

% Update last/next pointers in row r.
 NZ( r)=-A( r ,m);
 LZ( r)=j ;

% Mark A( r ,m) as an assigned zero in
%the matrix
 A( r ,m)=0;

% ...and in the assignment vector.
 C(m)=r ;

%Remove A( i , j) from unassigned list.
 A( i , lj)=A( i , j );

%Update last/next pointers in row r.
 NZ( i)=-A( i , j );
 LZ( i)=lj ;

% Mark A( r ,m) as an assigned zero in the
% matrix
 A( i , j )=0;

% ...and in the assignment vector.
 C( j)=i ;

% Stop search.
 break ;
```

```
end
end
end
end
```

```
%Create vector with list of unassigned
%rows.
```

```
% Mark all rows have assignment.
r=zeros(1,n);
rows=C(C~=0);
r(rows)=rows;
empty=find(r==0);
```

```
% Create vector with linked list of
%unassigned rows.
U=zeros(1,n+1);
U([n+1 empty])=[empty 0];
```

```
function [A,C,U]=hmflip(A,C,LC,LR,U,l,r)
%HMFLIP Flip assignment state of all
%zeros along a path.
%
%[A,C,U]=hmflip(A,C,LC,LR,U,l,r)
%Input:
%A - the cost matrix.
%C - the assignment vector.
%LC- the column label vector.
%LR- the row label vector.
%r,l- position of last zero in path.
%Output:
%A -updated cost matrix.
%C -updated assignment vector.
%U -updated unassigned row list vector

n=size(A,1);
```

```
while (1)
```

```
% Move assignment in column l to row r
C(l)=r;

% Find zero to be removed from zero list

% Find zero before this
m=find(A(r,:)==-1);

% Link past this zero
A(r,m)=A(r,l);

A(r,l)=0;

% If this was the first zero of the path
if (LR(r)<0)
%remove row from unassigned row list
%and return
U(n+1)=U(r);
U(r)=0;
return;
else

% Move back in this row along the path and
% get column of next zero
l=LR(r);

% Insert zero at (r,l) first in zero list
A(r,l)=A(r,n+1);
A(r,n+1)=-1;

% Continue back along the column to get
%row of next zero in path.
r=LC(l);
end
end


function [A,CH,RH]=hmreduce(A,CH,RH,LC,LR,
                                    SLC,SLR)
```

```
%Reduce parts of cost matrix in
% the Hungarian method.
%[A,CH,RH]=hmreduce(A,CH,RH,LC,LR,SLC,SLR)
%Input:
%A -Cost matrix.
%CH-vector of column of 'next zeros' in
%each row
%RH- vector with list of unexplored rows
%LC -column labels
%RC -row labels
%SLC-set of column labels
%SLR-set of row labels
%
%Output:
%A -Reduced cost matrix.
%CH-Updated vector of 'next zeros' in
%each row
%RH -Updated vector of unexplored rows


n=size(A,1);

%Find which rows are covered,
%i.e. unlabelled
coveredRows=LR==0;

%Find which columns are covered,
%i.e. labelled
coveredCols=LC~=0;

r=find(~coveredRows);
c=find(~coveredCols);

% Get minimum of uncovered elements.
m=min(min(A(r,c)));

%Subtract minimum from all uncovered elements
A(r,c)=A(r,c)-m;
```

```
% Check all uncovered columns
for j=c
% ... and uncovered rows in path order
for i=SLR
% If this is a (new) zero
if (A(i,j)==0)
% If the row is not in unexplored list
if (RH(i)==0)
% ... insert it first in unexplored list
RH(i)=RH(n+1);
RH(n+1)=i;
%Mark this zero as "next free" in this row
CH(i)=j;
end
% Find last unassigned zero on row I
row=A(i,:);
colsInList=-row(row<0);
if (length(colsInList)==0)
% No zeros in the list.
l=n+1;
else
l=colsInList(row(colsInList)==0);
end
% Append this zero to end of list.
A(i,l)=-j;
end
end
end

% Add minimum to all doubly covered elements.
r=find(coveredRows);
c=find(coveredCols);
% Take care of the zeros we will remove.
[i,j]=find(A(r,c)<=0);
i=r(i);
j=c(j);
for k=1:length(i)
% Find zero before this in this row.
lj=find(A(i(k),:)==-j(k));
```

```
% Link  past  it .
A( i ( k ) , l j )=A( i ( k ) , j ( k ) ) ;
% Mark  it  as  assigned .
A( i ( k ) , j ( k ))=0;
end

A( r , c )=A( r , c )+m;
```

Example 9.6. Solve the following assignment problem in MAT-LAB.

6768	124	916	2489
124	6768	6768	2489
2489	337	2489	6768
916	916	6768	10000

In the Command prompt

```
>> A=[6768 124 916 2489; 124 6768 6768 2489; 2489 337
2489 6768; 916 916 6768 10000]

>> [C,T]=hungarian(A)
```

Output:

```
C= 4  3  1  2

T= 4658
```

Example 9.7. Determine an optimal assignment in MATLAB for the following assignment problem.

12	16	14	10	5	12	18	13
8	8	7	9	8	11	10	12
13	18	16	20	9	11	14	17
14	18	17	19	12	10	15	14
13	15	12	13	6	18	13	13
6	5	8	9	8	7	4	7
1	4	7	6	3	3	2	5
11	9	10	12	7	5	7	11

In the Command prompt

```
>> A=[12 16 14 10 5 12 18 13; 8 8 7 9 8 11 10 12; 13
18 16 20 9 11 14 17; 14 18 17 19 12 10 15 14; 13 15
12 13 6 18 13 13; 6 5 8 9 8 7 4 7; 1 4 7 6 3 3 2 5;
11 9 10 12 7 5 7 11]

>> [C,T]=hungarian(A)
```

Output:

```
C= 7  6  2  1  5  3  8  4
T= 61
```

9.3   Exercises

Exercise 9.1. A traveling company owns cars in each of the five locations $L1$, $L2$, $L3$, $L4$, $L5$ and the passengers are in each of the five villages $V1$, $V2$, $V3$, $V4$, $V5$, respectively. The following table shows the distance between the locations and villages in kilometers. How should cars be assigned to the passengers so as to minimize the total distance covered?

	V1	V2	V3	V4	V5
L1	120	110	115	30	36
L2	125	100	95	30	16
L3	145	90	135	60	70
L4	160	140	150	60	60
L5	190	155	165	90	85

Village (column header); *Loc* (row header on left of L3)

Exercise 9.2. Solve the following minimal assignment problem.

	1	2	3	4	5	6
A	31	62	29	42	15	41
B	12	19	39	55	71	40
C	17	29	50	40	22	22
D	35	41	38	42	27	33
E	19	30	29	16	20	23
F	72	30	30	50	41	20

Man (column header); *Job* (row header on left of C)

Exercise 9.3. Five professors reach the Varanasi railway station in their hometown and want to travel to their respective homes in auto rickshaws. Each professor approaches a rickshaw driver and finds out the charge for the final destinations from him. The following table denotes the charges.

Loc

	1	2	3	4	5
A	60	90	40	60	40
B	30	70	50	80	50
C	40	90	70	60	60
D	80	50	60	40	50
E	70	80	60	60	60

Professor

Find out the total charge if the professors use the optimal assignment solution.

Exercise 9.4. An optimization class contains four students available for work on four projects. Only one student can work on any one project. The following table shows the cost of assigning each student to each project. The objective is to assign students to projects such that the total assignment cost is a minimum.

Student

	1	2	3	4
A	20	25	22	28
B	15	18	23	17
C	19	17	21	24
D	25	23	24	24

Project

Exercise 9.5. Solve the assignment problem represented by the following matrix which gives the distances from customers A, B, C, D, E to depots a, b, c, d, and e. Each depot has one car. How should the cars be assigned to the customers so as to minimize the distance travelled?

City

	a	b	c	d	e
A	160	130	175	190	200
B	135	120	150	160	175
C	50	50	180	180	110
D	160	140	130	60	60
E	55	35	80	80	105

Loc is at left, spanning C and D.

Exercise 9.6. Determine an optimal assignment for the assignment problem with the following rating matrix.

	A	B	C	D	E
1	7	9	10	3	7
2	5	10	11	6	6
3	2	6	9	1	8
4	9	8	7	5	9
5	3	8	6	4	8

Exercise 9.7. Find the optimal assignment for the following assignment problem given by the cost matrix.

	A	B	C	D	E
C1	11	14	8	16	20
C2	9	7	12	12	15
C3	13	16	15	6	16
C4	21	24	17	28	26
C5	17	10	12	11	15

Exercise 9.8. A software company has four system analysts, and four tasks have to be performed. System analysts differ in efficiency, and tasks differ in their intrinsic difficulty. The time that

each analyst would take to complete each task is given in the effectiveness matrix. How should the tasks be allocated to each analyst so as to minimize the total man-hours?

Task

		1	2	3	4
	A	5	23	14	8
Analyst	B	10	15	1	23
	C	35	16	12	15
	D	16	23	21	7

Exercise 9.9. Find the optimal assignment profit from the following matrices.

(a)

	1	2	3	4
A	7	5	4	3
B	8	2	6	4
C	5	3	2	1
D	5	4	1	8

(b)

	A	B	C	D	E
1	32	38	40	28	40
2	40	24	28	21	36
3	41	127	33	30	37
4	22	38	41	36	36
5	29	33	40	35	39

Exercise 9.10. Solve the minimal assignment problem.

$$Man$$

	M1	M2	M3	M4	M5
J1	7	17	8	16	20
J2	9	11	12	6	15
J3	13	16	15	12	16
J4	21	25	17	28	26
J5	14	10	12	11	15

Job appears to the left of J3/J4.

Exercise 9.11. Solve the assignment problem.

	1	2	3	4	5
A	6	5	8	11	16
B	1	16	16	1	10
C	13	11	8	8	8
D	9	14	12	10	16
E	10	13	11	8	16

Answer Key

Chapter 2

Exercise 2.1 (a) Linearly dependent (b) Linearly independent (c) Linearly independent. Exercise 2.2 (a) Rank=3 (b) Rank=3. Exercise 2.5 (a) Unique solution x=2, y=−2, z=2 (b) Infinitely many solutions.

Chapter 3

Exercise 3.1 (a) 2 (b) $1/50$ (c) –8. Exercise 3.2 (a) $\begin{bmatrix} 2 & 2 \\ 2 & 2 \end{bmatrix}$ (b) $\begin{bmatrix} 1 & 2 \end{bmatrix}$ (c) $\begin{bmatrix} -3 & 0 \\ 0 & 1 \end{bmatrix}$ (d) $\begin{bmatrix} 3 & 0 \\ 2 & 1 \end{bmatrix}$. Exercise 3.3 (a) 4, 5 (b) 5 (c) 4, 3 (d) 2, 1. Exercise 3.4 (a) 1 (b) 0 (c) 0 (d) 0 (e) 2 (f) 1 (g) 0

Chapter 4

Exercise 4.1 $x_1 = 5/2$, $x_2 = 35$, max=295/2. Exercise 4.2 $x_1 = 160$, $x_2 = 0$, max=4800. Exercise 4.3 (a) nonconvex set (b) convex set (c) nonconvex set (d) convex set (e) nonconvex set. Exercise 4.4 (a) $x_1 = 8$, $x_2 = 0$, max=23. (b) $x_1 = 0$, $x_2 = 1$, max=3. (c) no feasible corner (d) unbounded (e) $x_1 = 11$, $x_2 = 35$, max=323000. Exercise 4.5 (a) $x_1 = 2$, $x_2=1/2$, max=12 (b) $x_1 = 0$, $x_2 = 3$, max=6. Exercise 4.6 $x_1 = 5$, $x_2 = 6$, max=1100.

Chapter 5

Exercise 5.1 (a) $\begin{bmatrix} 0 & 20/3 & 37/3 & 0 \end{bmatrix}^T$ and $\begin{bmatrix} 20/3 & 0 & 17/3 & 0 \end{bmatrix}^T$ basic feasible (b) basic but no feasible. Exercise 5.3 $x_1=13/57$, $x_3=49/57$, max=928/57. Exercise 5.4 $x_1=23/10$, $x_2=1/2$, min=303/10. Exercise 5.5 (a) $x_1=7/3$, $x_2=2/3$, max=5/3 (b) $x_1=1/3$, $x_2=8/3$, max=7/3. Exercise 5.6 $x_1=3$, max=21. Exercise

5.7 $x_1=5$, $x_2=1$, max=52. Exercise 5.8 $x_2=15/2$, max=45. Exercise 5.9 $x_1 = 3/2$, $x_2 = 0$, max=3. Exercise 5.10 unbounded. Exercise 5.11 $x_3 = 7/6$, $x_4 = 1/3$, $x_5 = 13/6$, min=$\frac{58}{3}$. Exercise 5.12 $x_1 = 5$, $x_2 = 0$, min=25.

Chapter 6

Exercise 6.1 unbounded. Exercise 6.2 $x_1 = 0$, $x_2 = 0$, $x_3 = 6$, max=6. Exercise 6.3 $x_2 = 1$, $x_3 = 4$, max=13. Exercise 6.4 $x_1 = 21$, $x_2 =6$, max=549. Exercise 6.5 (a) $x_1 = 8/7$, $x_2 = 24/7$, max=$104/7$ (b) $x_1 = 8/7$, $x_2 = 24/7$, max=$104/7$. Exercise 6.6 $x_1 = 4$, $x_2 = 2$, max=24. Exercise 6.7 $x_1 = x_3 = 0$, $x_2 = 7/4$, max=14. Exercise 6.8 $x_1 = 39$, $x_2 = 0$, $x_3 = 48$, $x_4 = 30$, max=1827. Exercise 6.9 (a) $x_1 = 0$, $x_2 = 2$, max=6. (b) $x_1 = 2/3$, $x_2 = 4/3$, max=$34/3$. Exercise 6.10 $x_1 = 0$, $x_2 = 11$, max=33. Exercise 6.11 $x_1 = 0$, $x_2 = 7$, max=28.

Chapter 7

Exercise 7.1 (a) min $100y_1+90y_2+400y_3$ s.t. $5y_1 + y_2 \geq 10$, $-4y_1 + 12y_2 + y_3 \geq 30$ (b) max $-30y_1 - 50y_2 - 70y_3$ s.t. $6y_1 - 2y_2 \leq 3$, $11y_1 + 7y_2 - y_3 \leq -4$ (c) min $60y_1 - 10y_2 - 20y_3 + 20y_4$ s.t. $5y_1 - 3y_2 - y_3 + y_4 \geq 3$, $y_1 + 8y_2 - 7y_3 + 7y_4 \geq 2$. Exercise 7.3 $x_1=3/5$, $x_2=6/5$, min=12/5. Exercise 7.4 $x_1=25/22$, $x_2=9/11$,$x_3=27/22$, min=458/11. Exercise 7.5 $x_2=4$, $x_4=20$, min=168. Exercise 7.6 $x_1=2$, $x_2=1$, min=7. Exercise 7.7 (a) $x_1=11/7$, $x_2=1/7$, min=37/7. Exercise 7.8 $x_1 = 5, x_2 = 3, x_3 = 0$ min=106. Exercise 7.9 $x_1 = 20/3, x_2 = 0, x_3 = 50/3$, min=1900/3. Exercise 7.10 $x_1 = 6, x_2 = 21$, min=144. Exercise 7.11 $x_2 = 34, x_4 = 80$, min=182. Exercise 7.12 $x_2 = 86/9, x_3 = 10/9$, min=$\frac{719}{9}$. Exercise 7.13 $x_1 = x_2 = 1/3$, min=$2/3$, $x_1 = 1/2, x_4 = 1/4$, min=$7/12$.

Chapter 8

Exercise 8.1 (a) $x_{11}=2$,$x_{21}=1$, $x_{22}=3$, $x_{23}=2$, $x_{33}=2$, $x_{34}=5$, minT-cost= 204 (b) $x_{11}=18$, $x_{12}=4$, $x_{22}=12$, $x_{32}=3$,$x_{33}=4$, $x_{43}=4$, $x_{44}=2$,minTcost = 279. Exercise 8.2 (a) $x_{21}=20$, $x_{24}=55$, $x_{34}=5$, $x_{12}=20$, $x_{13}=30$, $x_{33} = 20$, minTcost=650 (b) $x_{33} = 7$, $x_{41} = 7$, $x_{43} = 7$, $x_{13} = 4$, $x_{22} = 8$, $x_{12} = 1$, minTcost=83. Exercise 8.3(a) $x_{24} = 25$, $x_{21} = 12$, $x_{12} = 18$, $x_{31} = 4$, $x_{13} = 1$, $x_{33} = 30$, minT-cost=355 (b) $x_{11} = 5$, $x_{43} = 14$, $x_{21} = 2$, $x_{22} = 6$, $x_{32} = 3$, $x_{33} = 4$, minTcost=102. Exercise 8.4 (a) $x_{11} = 5$, $x_{12} = 3$, $x_{22} = 8$,

$x_{23} = 5$, $x_{33} = 3$, $x_{34} = 8$, minTcost=486 (b) $x_{24} = 8$, $x_{11} = 5$, $x_{13} = 3$, $x_{33} = 5$, $x_{32} = 6$, $x_{22} = 5$ minTcost=366 (c) $x_{32} = 11$, $x_{24} = 8$, $x_{11} = 5$, $x_{13} = 3$, $x_{23} = 5$, minTcost=366. Exercise 8.5 optimum cost=180. Exercise 8.6 optimal cost=80. Exercise 8.7 optimum cost=743. Exercise 8.8 optimal cost=723.

Chapter 9

Exercise 9.1 L1→C1, L2→C3, L3→C2, L4→C4, L5→C5, 450. Exercise 9.2 A→5, B→2, C→1, D→3, E→4, F→6, 125. Exercise 9.3 A→5, B→1, C→4, D→2, E→3, 240. Exercise 9.4 A→3, B→1, C→2, D→4, 78. Exercise 9.5 $A → b$, $B → c$, $C → a$, $D → d$, $E → e$, 470. Exercise 9.6 $1 → D$, $2 → E$, $3 → A$, $4 → B$, $5 → C$,25. Exercise 9.7 $C1 → A$, $C2 → B$, $C3 → D$, $C4 → C$, $C5 → E$, 56. Exercise 9.8 $A → 1$, $C → 2$, $B → 3$, $D → 4$, 29. Exercise 9.9 (a)$A → 1$, $B → 2$, $C → 4$, $D → 3$, 11 (b) $4 → A$, $2 → B$, $3 → C$, $1 → D$, $5 → E$, 146. Exercise 9.10 J1→M1, J2→M4, J3→M5, J4→M3, J5→M2, 56. Exercise 9.11 1→B, 2→A, 3→D, 4→E, 5→C, 34.

Bibliography

[1] E. K. P. Chong and S. H. Zak. *An Introduction to Optimization*. John Wiley & Sons, Inc, New Jersey, USA, 2013.

[2] G. Giorgi and T. H. Kjeldsen. *Traces and Emergence of Nonlinear Programming*. Springer, Basel, 2014.

[3] B. R. Hunt, R. L. Lipsman, and J. M. Rosenberg. *A Guide to MATLAB for Beginners and Experienced Users*. Cambridge University Press, USA, 2006.

[4] V. Krishnamurthy, V. P. Mainra, and J. L. Arora. *An Introduction to Linear Algebra*. East West Press Ltd. New Delhi, 1976.

[5] D. G. Luenberger. *Linear and Nonlinear Programming*. Addison Wesley Publishing Company, USA, 1984.

[6] B. N. Mishra and B. K. Mishra. *Optimization: Linear Programming*. Ane Books Pvt. Ltd, New Delhi, 2006.

[7] P. K. Singh. *Math-Level Multiobjective Planning for Agriculture for District Mau, UP*. PhD thesis, Institute of Agricultural Sciences, Banaras Hindu University, Varanasi, 2003.

[8] A. Sultan. *Linear Programming-An Introduction with Applications*. Academic Press, Inc., USA, 1993.

[9] S. Vajda. *Problems in Linear and Non-Linear Programming*. Charles Griffin and Company Ltd, London and High Wycombe, 1975.

[10] S. Vajda. *Linear Programming-Algorithms and Applications*. Chapman and Hall Ltd, New York, USA, 1981.

[11] S. Vajda. *Mathematical Programming*. Dover Publications Inc., New York, 1989.

Index

abs, 27
Algorithm
  The Revised Simplex
    Algorithm, 139
artificial problem, 119
assignment problem, 273
asymmetric form of duality, 177
augmented matrix, 14

Balanced Transportation
    Problem, 215
bascicell6.m, 262
basic feasible solution, 84
basic solution, 70
BFS, 249
built-in, 25

canonical augmented matrix, 87
Cleve Barry Moler, 4
Command Window, 21
comments, 27
complementary slackness
    condition, 183
computational procedure, 249
convex set, 50
cycle5.m, 260

degenerate, 217
degenerate basic solution, 70
demand center, 219
Dual Linear Programs, 175
dual of the dual, 176
Dual Simplex Algorithm

Algorithm, 185
Dual Simplex Method, 185
dual.m, 190
Duality Theorem, 181

elementary row operation, 11
end, 26
extreme points, 53, 81
eye, 25

feasible point, 45
feasible solution, 70, 80, 217
fictitious source center, 220
floating point, 30
fprintf(), 29
Fundamental Theorem of
    Linear Programming
    Problem, 78

George B. Dantzig, 3

hotel requirement, 48
Hungarian method, 275

if, 32
inequality, 46
input, 28
investment of funds, 56

Jeno Egervary, 275
Joseph B. Fourier, 1
Julius Farkas, 2

Kantorovich, 3